Norio Baba, Lakhmi C. Jain and Hisashi Handa (Eds.)

Advanced Intelligent Paradigms in Computer Games

Studies in Computational Intelligence, Volume 71

Editor-in-chief
Prof. Janusz Kacprzyk
Systems Research Institute
Polish Academy of Sciences
ul. Newelska 6
01-447 Warsaw
Poland
E-mail: kacprzyk@ibspan.waw.pl

Further volumes of this series
can be found on our homepage:
springer.com

Vol. 50. Nadia Nedjah, Leandro dos Santos Coelho,
Luiza de Macedo Mourelle (Eds.)
Mobile Robots: The Evolutionary Approach, 2007
ISBN 978-3-540-49719-6

Vol. 51. Shengxiang Yang, Yew Soon Ong, Yaochu Jin
Honda (Eds.)
*Evolutionary Computation in Dynamic and Uncertain
Environment*, 2007
ISBN 978-3-540-49772-1

Vol. 52. Abraham Kandel, Horst Bunke, Mark Last (Eds.)
*Applied Graph Theory in Computer Vision and Pattern
Recognition*, 2007
ISBN 978-3-540-68019-2

Vol. 53. Huajin Tang, Kay Chen Tan, Zhang Yi
*Neural Networks: Computational Models
and Applications*, 2007
ISBN 978-3-540-69225-6

Vol. 54. Fernando G. Lobo, Cláudio F. Lima
and Zbigniew Michalewicz (Eds.)
Parameter Setting in Evolutionary Algorithms, 2007
ISBN 978-3-540-69431-1

Vol. 55. Xianyi Zeng, Yi Li, Da Ruan and Ludovic Koehl
(Eds.)
Computational Textile, 2007
ISBN 978-3-540-70656-4

Vol. 56. Akira Namatame, Satoshi Kurihara and
Hideyuki Nakashima (Eds.)
Emergent Intelligence of Networked Agents, 2007
ISBN 978-3-540-71073-8

Vol. 57. Nadia Nedjah, Ajith Abraham and Luiza de
Macedo Mourella (Eds.)
*Computational Intelligence in Information Assurance
and Security*, 2007
ISBN 978-3-540-71077-6

Vol. 58. Jeng-Shyang Pan, Hsiang-Cheh Huang, Lakhmi
C. Jain and Wai-Chi Fang (Eds.)
Intelligent Multimedia Data Hiding, 2007
ISBN 978-3-540-71168-1

Vol. 59. Andrzej P. Wierzbicki and Yoshiteru
Nakamori (Eds.)
Creative Environments, 2007
ISBN 978-3-540-71466-8

Vol. 60. Vladimir G. Ivancevic and Tijana T. Ivacevic
*Computational Mind: A Complex Dynamics
Perspective*, 2007
ISBN 978-3-540-71465-1

Vol. 61. Jacques Teller, John R. Lee and Catherine
Roussey (Eds.)
Ontologies for Urban Development, 2007
ISBN 978-3-540-71975-5

Vol. 62. Lakhmi C. Jain, Raymond A. Tedman
and Debra K. Tedman (Eds.)
*Evolution of Teaching and Learning Paradigms
in Intelligent Environment*, 2007
ISBN 978-3-540-71973-1

Vol. 63. Wlodzislaw Duch and Jacek Mańdziuk (Eds.)
Challenges for Computational Intelligence, 2007
ISBN 978-3-540-71983-0

Vol. 64. Lorenzo Magnani and Ping Li (Eds.)
*Model-Based Reasoning in Science, Technology, and
Medicine*, 2007
ISBN 978-3-540-71985-4

Vol. 65. S. Vaidya, L.C. Jain and H. Yoshida (Eds.)
*Advanced Computational Intelligence Paradigms in
Healthcare-2*, 2007
ISBN 978-3-540-72374-5

Vol. 66. Lakhmi C. Jain, Vasile Palade and Dipti
Srinivasan (Eds.)
*Advances in Evolutionary Computing for System
Design*, 2007
ISBN 978-3-540-72376-9

Vol. 67. Vassilis G. Kaburlasos and Gerhard X. Ritter
(Eds.)
*Computational Intelligence Based on Lattice
Theory*, 2007
ISBN 978-3-540-72686-9

Vol. 68. Cipriano Galindo, Juan-Antonio
Fernández-Madrigal and Javier Gonzalez
*A Multi-Hierarchical Symbolic Model
of the Environment for Improving Mobile Robot
Operation*, 2007
ISBN 978-3-540-72688-3

Vol. 69. Falko Dressler and Iacopo Carreras (Eds.)
*Advances in Biologically Inspired Information Systems:
Models, Methods, and Tools*, 2007
ISBN 978-3-540-72692-0

Vol. 70. Javaan Singh Chahl, Lakhmi C. Jain, Akiko
Mizutani and Mika Sato-Ilic (Eds.)
Innovations in Intelligent Machines-1, 2007
ISBN 978-3-540-72695-1

Vol. 71. Norio Baba, Lakhmi C. Jain and Hisashi Handa
(Eds.)
*Advanced Intelligent Paradigms in Computer
Games*, 2007
ISBN 978-3-540-72704-0

Norio Baba
Lakhmi C. Jain
Hisashi Handa
(Eds.)

Advanced Intelligent Paradigms in Computer Games

With 43 Figures and 16 Tables

Springer

Norio Baba
Osaka Kyoiku University
Department of Information Science
Asahiga-oka
4-698-1 Kashiwara
Osaka 582-8582
Japan
E-mail:- baba@is.osaka-kyoiku.ac.jp

Hisashi Handa
Graduate School of Natural Science
& Technology
Division of Industrial Innovation Sciences
Okayama University
Tsushima-Naka 3-1-1
Okayama 700-8530
Japan
E-mail:- handa@sdc.it.okayama-u.ac.jp

Lakhmi C. Jain
University of South Australia
Mawson Lakes Campus
Adelaide, South Australia
Australia
E-mail:- Lakhmi.jain@unisa.edu.au

Library of Congress Control Number: 2007927410

ISSN print edition: 1860-949X
ISSN electronic edition: 1860-9503
ISBN 978-3-540-72704-0 Springer Berlin Heidelberg New York

This work is subject to copyright. All rights are reserved, whether the whole or part of the material is concerned, specifically the rights of translation, reprinting, reuse of illustrations, recitation, broadcasting, reproduction on microfilm or in any other way, and storage in data banks. Duplication of this publication or parts thereof is permitted only under the provisions of the German Copyright Law of September 9, 1965, in its current version, and permission for use must always be obtained from Springer-Verlag. Violations are liable to prosecution under the German Copyright Law.

Springer is a part of Springer Science+Business Media
springer.com
© Springer-Verlag Berlin Heidelberg 2007

The use of general descriptive names, registered names, trademarks, etc. in this publication does not imply, even in the absence of a specific statement, that such names are exempt from the relevant protective laws and regulations and therefore free for general use.

Cover design: deblik, Berlin
Typesetting by the SPi using a Springer LATEX macro package
Printed on acid-free paper   SPIN: 11615316   89/SPi   5 4 3 2 1 0

Preface

The evolution of technologies has greatly changed the basic structure of our industry and nature of our daily lives. Industries which did not exist several decades ago have made remarkable progress in recent years and flourished. One of the most typical examples is the computer game industry. This book presents a sample of the most recent research concerning the application of computational intelligence techniques and internet technology in computer games.

This book contains eight chapters. The first chapter, by N. Baba and H. Handa, is on utilization of evolutionary algorithms to increase excitement of the COMMONS GAME. It is shown that the original COMMONS GAME which is one of the most popular environmental games has been made much more exciting by the intelligent utilization of the two evolutionary algorithms.

The second chapter, by H. Barber and D. Kudenko, is on adaptive generation of dilemma-based interactive narratives. In this chapter, they present an interactive narrative generator that can create story lines that incorporate dilemmas to add dramatic tension. They also briefly touch upon the possibility that their work could provide a useful tool for making dramatically interesting game playing possible.

The third chapter, by J. Tongelius, S.M. Lucas, and R.D. Nardi, is on computational intelligence (CI) in racing games. The authors suggest that CI techniques can be used for various purposes such as controller evolution for the racing and track evolution for a proficient player.

In the fourth chapter, K.J. Kim and S.B. Cho present several methods to incorporate domain knowledge into evolutionary board game players. Several experimental results concerning game playing of Checkers and Othello confirm that their approach can improve the performance of evolved strategies compared to that without knowledge.

The fifth chapter by C. Frayn and C. Justiniano is on the ChessBrain project – massively distributed chess tree search. This chapter presents the ChessBrain project which has investigated the feasibility of massively parallel game tree search using a diverse network of volunteers connected via the

Internet. It also presents their recent project named ChessBrain II which is aims to reduce the inefficiencies experienced by its predecessor ChessBrain I.

In the sixth chapter, M. Fasli and M. Michalakopoulos present the e-Game which enables both web users and software agents to participate in electronic auctions. They also present the future possibility for multi-attribute auctions where behaviors of some participated agents could be controlled using the computational intelligence techniques.

The seventh chapter, by K. Burns, is on EVE's entropy: a formal gauge of fun in games. In this chapter, the author suggests that EVE's theory could contribute a lot for constructing a new paradigm to analyze fun in gambling games.

The last chapter, by G.N. Yannakakis and J. Hallam, is on capturing player enjoyment in computer games. This chapter presents two dissimilar approaches for modeling player's satisfaction in computer games. They utilize the familiar game Pac-Man for investigating the quantitative impact of the factors on player's entertainment.

This book will prove useful to researchers, practicing engineers/scientists, and students who are interested to learn about the state-of-the-art technology.

We would like to express our sincere gratitude to the authors and reviewers for their wonderful contributions and vision.

Editors

Contents

COMMONS GAME Made More Exciting by an Intelligent Utilization of the Two Evolutionary Algorithms
Norio Baba, Hisashi Handa 1

Adaptive Generation of Dilemma-based Interactive Narratives
Heather Barber, Daniel Kudenko 19

Computational Intelligence in Racing Games
Julian Togelius, Simon M. Lucas and Renzo De Nardi 39

Evolutionary Algorithms for Board Game Players with Domain Knowledge
Kyung-Joong Kim and Sung-Bae Cho 71

The ChessBrain Project – Massively Distributed Chess Tree Search
Colin Frayn, Carlos Justiniano 91

Designing and Developing Electronic Market Games
Maria Fasli and Michael Michalakopoulos 117

EVE's Entropy: A Formal Gauge of Fun in Games
Kevin Burns 153

Capturing Player Enjoyment in Computer Games
Georgios N. Yannakakis, John Hallam 175

COMMONS GAME Made More Exciting by an Intelligent Utilization of the Two Evolutionary Algorithms

Norio Baba[1] and Hisashi Handa[2]

[1] Department of Information Science, Osaka Kyoiku University
Kashihara 4-698-1, Osaka Prefecture, 582-8582, JAPAN
baba@cc.osaka-kyoiku.ac.jp
[2] Graduate School of Natural Science and Technology, Okayama University
Tsushima-Naka 3-1-1, Okayama 700-8530, JAPAN
handa@sdc.it.okayama-u.ac.jp

In this paper, we suggest that Evolutionary Algorithms could be utilized in order to let the COMMONS GAME, one of the most popular environmental games, become much more exciting. In order to attain this objective, we utilize Multi-Objective Evolutionary Algorithms to generate various skilled players. Further, we suggest that Evolutionary Programming could be utilized to find out an appropriate point of each card at the COMMONS GAME. Several game playings utilizing the new rule of the COMMONS GAME confirm the effectiveness of our approach.

1 Introduction

Gaming is regarded by many people as a new and promising tool to deal with complex problems in which human decisions have far reaching effects on others. It has been used for various purposes such as decision-making, education, training, research, entertainment, and etc. [1]-[12]. In recent years, various approaches concerning the applications of Evolutionary Algorithms to the field of games have been proposed [13]-[17].

In this paper, we suggest that EAs could be utilized for making the COMMONS GAME [8], one of the most popular environmental games, become much more exciting. In particular, in order to attain this objective, we shall try to utilize Evolutionary Algorithms in the following steps:

1) First, we shall consider a new rule for assigning a point to each colored card in the COMMONS GAME which takes the environmental changes into account.

2) Second, we shall utilize Multi-Objective Evolutionary Algorithms (MOEA) [18][19] to generate various skilled players whose choice of each card is done in a timely fashion.
3) Further, we shall utilize Evolutionary Programming (EP) [20][21] to derive appropriate combinations of the rules (concerning the point of each card) which could be used to help players fully enjoy game playing.

This paper is organized as follows. In section 2, we shall introduce the original COMMONS GAME briefly and touch upon several problems involved in the original COMMONS GAME. We shall suggest that EAs could be utilized in order to let game playing become much more exciting. We shall also show several results of game playing (utilizing the new rule derived by MOEA & FEP) which confirm the effectiveness of our approach. This paper concludes with discussions concerning the contributions of this paper and future perspectives.

2 COMMONS GAME

2.1 History of Gaming

Historically speaking, gaming[1] has its origin in war games [4]. However, after the Second World War, it has been applied to various peaceful purposes. A large number of business games have been developed with the purpose of training students in business school [2][5][6]. Further, some environmental games have also been developed in order to help people consider seriously about the environmental state of the world [8]-[10]. Gaming has also successfully been utilized for operational purposes [7]. Depending upon the purpose of the game, gaming can be categorized into several classes such as Entertainment Gaming, Educational Gaming, Operational Gaming and etc. [1][2] Due to space, we don't go into details concerning the literature of gaming and the categorization of gaming. Interested readers are kindly asked to read the books and papers [1]-[12].

In the following subsection, we shall briefly introduce COMMONS GAME[2] [8]. We shall also briefly touch upon the computer gaming system of the COMMONS GAME.

2.2 Brief Introduction of the COMMONS GAME

The COMMONS GAME was developed by Powers *et al.* in 1977 [8]. Since we live in a world having only finite natural resources such as fishes and forests

---
[1] Occasionally, game theory has been confused with gaming. Gaming means the use of a game for one of the various purposes such as teaching, training, operations, entertainment, and etc. [1][2].
[2] The COMMONS GAME was designed by Powers *et al.* [8] in order to let people have a chance to consider seriously about the COMMONS. Therefore, COMMONS GAME can be categorized into the class of the Educational Gaming [1].

(commons), it is wise to consider their careful utilization. The COMMONS GAME may be quite helpful in stimulating discussion of this problem. Figure 1 shows the layout of the original COMMONS GAME.

In the following, we give a brief introduction to this game. Six players are asked to sit around a table. Following a brief introduction of the game, the game director tells the players that their objective is to maximize their own gains by choosing one card among the five colored (Green, Red, Black, Orange, Yellow) cards in each round. In each round, players hide their cards behind a cardboard shield to ensure individual privacy.

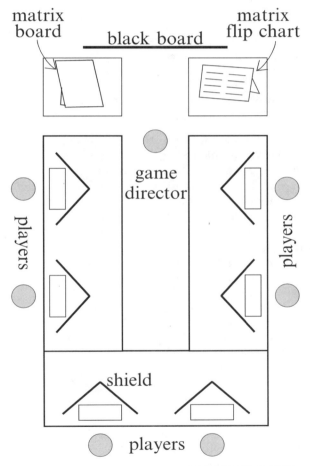

Fig. 1. Layout of the original COMMONS GAME

Each colored card has its own special meaning concerning the attitude toward the environmental protection, and has the following effect upon the total gains of each player:

Green card: A green card represents high exploitation of the commons: Players who play a green card can get a maximum reward. However, they lose 20 points if one of the other players plays a black card in the same round.

Red card: A red card indicates a careful utilization of the commons: Red cards are only worth about forty percent as many points as green cards.

Black card: This card has a punishing effect on the players with green cards: Players who have played a black card have to lose 6 points divided by the number of black cards played at the same round, but are able to punish green card players by giving them -20 points.

Orange card: An orange card gives an encouraging effect to red card players: Players who have played this card have to lose 6 points divided by the number of orange cards played at the same round but are able to add 10 points to red card players.

Yellow card: A yellow card denotes a complete abstention from utilization of the commons: Players who play this card get 6 points.

Depending upon the players' strategies, the state of the commons change: If players are too eager to exploit the commons, then deterioration of the commons occurs. Players have a matrix flip chart on the board representing the payoffs for the red and green cards under different conditions of the commons. Although players are informed that there will be 60 rounds, each game ends after 50 rounds. After each 8th round, players have a three-minute conference. They can discuss everything about the game and decide every possible way to play in future rounds.

Due to space, the detail of the rules are not explained. For those interested in further details, however, we recommend reading the paper written by Powers *et al.* [8].

Remark 2.1. In each round, each player can choose one of the 5 cards. However, COMMONS GAME is quite different from traditional "card game" in which cards are distributed randomly to the players and losing cards is frequently occurred. In this game, each player can choose any card in each round in order to represent his (or her) attitude toward the environmental protection.

Remark 2.2. Red players usually can only get about 40 % of the returns that green card players receive (assuming that no players have chosen the black card at the same round). In the original COMMONS GAME, there are 17 main environmental states $(-8, -7, \ldots, -1, 0, +1, \ldots, +8)$. (Each main environmental state (except 0 state) has 10 subordinate states (the 0 state has 21 subordinate states)). Initial state of the COMMONS GAME is 0. The state of the commons changes, depending upon the players' strategies. If players

Table 1. Points which can be gained by the red and the green card players

| State: -8 || State: -1 || State: 0 || State:1 || State:8 ||
R	G	R	G	R	G	R	G	R	G
—	0	—	70	—	100	—	130	—	200
-10	2	25	72	40	102	55	132	90	202
-8	4	27	74	42	104	57	134	92	204
-6	6	29	76	44	106	59	136	94	206
-4	8	31	78	46	108	61	138	96	208
-2	10	33	80	48	110	63	140	98	210
0	—	35	—	50	—	65	—	100	—

are eager to exploit the commons (many players often use green cards), the deterioration of the commons occurs and the state of the commons changes to a minus state such as $-1, -2$, and etc. When the deterioration of the commons has occurred, then the returns that green players and red players receive decrease. Table 1 shows the returns that green players can get when no players have chosen the black card nor the orange card. It also shows the returns that red players can receive. The second table from the left shows that green players can receive 70 point (when the state of the environment is -1 and no players have chosen the red cards) which is only the 70 % of the returns which can be obtained at the initial state 0. This table also shows that the points that green players can get change depending upon the number of the red cards played in the same rounds (Remark 2.4). The points that red card players receive also decrease heavily when the environmental state becomes minus. In the -1 state, red players can get only about 70 % of the returns that they could receive in the initial state 0. (On the other hand,) If almost all of the players consider seriously about the commons and execute wise utilization of the commons, then environmental state ameliorates (state of the environmental becomes positive) and the returns that green players and red players receive increase as shown in Table 1.

Remark 2.3. Utilization of a green card also incurs degradation of the subordinate state. Although natural amelioration of the commons occurs 8 times during 50 rounds, too much exploitation of the commons (that is to say, too much use of the green card) causes serious degradation of the environmental state (One green card corresponds to one degradation of the subordinate state).

Remark 2.4. Each row in the five tables in Table 1 corresponds numbers of the red cards chosen. For an example, let us consider that case that the current main state is $+1$, and numbers of the red and green card players are 5 and 1, respectively. Then, each red card player can get point 63 which corresponds to the point written in the 6th row and the 1st column of the table concerning the state $+1$. The green card player can get point 140.

2.3 Computer Gaming System of the COMMONS GAME

More than 20 years ago, one of the authors and his students succeeded in constructing a personal computer gaming system of the original COMMONS game [11][12]. In this gaming system, each player does not need to play one of the five cards in order to show his decision concerning the commons. Instead, he has to choose a column and a row number in a matrix written on a paper delivered by the game director.

A computer screen gives players various information such as the state of the commons and points received by the players in each round. If the state of the commons declines, the color of the waves becomes tinged with yellow. Also, the color of the waves becomes tinged with blue if the state of the commons improves. During the conference time, the computer screen provides players with a beautiful color graphic and gives information regarding the time passed.

3 Evolutionary Algorithms for Making Game Playing Much More Exciting

We have so far enjoyed a large number of playings of the original COMMONS GAME. Although those experiences have given us a valuable chance to consider seriously about the commons, we did find that some players lost interest, in the middle of the game, because the COMMONS GAME is comparatively monotonous. In order to make the game much more exciting [22]-[29], we have tried to find the reason why some players lost interest in the middle of the COMMONS GAME. We have come to the conclusion that the way that each player receives points when he (or she) chooses one of the five cards sometimes makes the game playing rather monotonous.

In particular, we have concluded that the following rule make the game playing monotonous: In the original COMMONS GAME, green card players receive a penalty, -20 points, when some player chooses a black card. On the other hand, black card players receive a point $-6/$(the number of players who have chosen a black card). Orange card players receive a point $-6/$(the number of players who have chosen an orange card).

We consider that some change in the points -20 and -6 mentioned above would make the COMMONS GAME much more exciting. In order to find an appropriate point for each card, we shall try to utilize EP.

In section 3.1, we suggest that Multi-Objective Evolutionary Algorithms (MOEA) [18][19] can generate various kinds of Neural Network Players with different strategies. In section 3.2, we show that Evolutionary Programming [20][21] can be a useful tool for finding appropriate points of the cards in the COMMONS game.

3.1 MOEAs for Generating Intelligent Players

Multi-Objective optimization is one of the most promising fields in Evolutionary Algorithms research. Due to the population search of EAs, Multi-Objective Evolutionary Algorithms (MOEAs) can evolve candidates of Pareto optimal solutions. Hence, in comparison with conventional EAs, MOEAs can simultaneously find out various solutions. In this paper, we employ NSGA-II [18], proposed by Deb et al., to evolve game players with Neural Networks. The NSGA-II utilizing crowded tournament selection, the notion of archive, and ranking method with non-dominated sort, is one of the most famous MOEAs. Most recent studies proposing new MOEAs cite their paper [19]. (A brief introduction of the NSGA-II is given in the Appendix.)

Neural Network Model. Our objective is to simultaneously construct plenty of Neural Network players with different strategies. The neural network model adopted in this paper has 26 input units, one hidden layer with 30 units, and 5 output units[3].

In Table 2, input variables into this neural network model are given: In order to represent the current state of the commons, two inputs, consisting of main state and subordinate state, are prepared. For inputs 6.–9., 5 different inputs corresponding to each card are prepared.

The weights of the neural network model are evolved by MOEA. That is, the number of gene in an individual is 965 (the number of weights between input layer and hidden layer: 26×30, the number of weights between hidden layer and output layer: 30×5, and the number of thresholds: 5).

Fitness Evaluation. In order to evaluate each individual, it is better to let him play with various game players. Fitness evaluation of individuals is carried out as follows:

Table 2. Input variables for Neural Network players

1.	Difference between the total points of each player and the average
2.	Rank of each player
3.	States of the environment: main state & subordinate state
4.	Changes in the environment
5.	Round Number
6.	Weighted sum of each card having been chosen by all of the players
7.	Weighted sum of each card having been chosen by each player
8.	The number of each card chosen in the previous round
9.	The card chosen by each player in the previous round

---

[3] Each output unit corresponds to each colored card. The colored card corresponding to the output unit which has emitted the highest output value is considered to be that chosen by the neural network player.

Table 3. Efficiency of each card

Player's card	$E_i(C)$	Situations
R	+1	No black player, but some green players
	−1	Otherwise
B	+1	No green player
	−1	Otherwise
G	+1	Some black players
	−1	Otherwise

1) Choose 30 individuals randomly from the parent population at each generation, where they become the opponents for 6 game runs (5 individuals are needed as opponents for a single game run).
2) The child population is generated from the parent population.
3) Each individual in the parent and child populations plays with the opponents chosen in **1)**.
4) As a consequence of game runs, individuals are evaluated with two objective functions[4] O_v and O_e: Variance of the total number of each card chosen in each game run and the efficiency of the cards played, respectively. Variance of the number of the card played is calculated by the following equation:

$$O_v = V_{RGB} + 20 * N_{OY},$$

where V_{RGB} is the variance of the total number of red, green, and black cards played in each game run, and N_{OY} is the total number of orange and yellow cards chosen. The efficiency of the cards used is calculated by integrating the evaluation at each round (Table 3 shows the way how the efficiency of the cards played at each round is evaluated):

$$O_e = \sum_{i=1}^{50} E_i(C).$$

Experimental Results Figure 2 depicts the individual distributions at the initial and final generations. In the individual distributions at the final generation, a Pareto set is constructed. Since fitness measurement in this paper is a relative one, i.e., opponents are randomly chosen at every generation, some initial individuals in which the efficiency of the use of the cards and the variance are close to −48 and 400, respectively, seem to have better performance. However, it is just caused by the game playings with naive players. In fact, the final individuals have become much more sophisticated compared with all of the initial individuals.

---

[4] According to the implementation of NSGA-II, the objective functions used here are to be minimized.

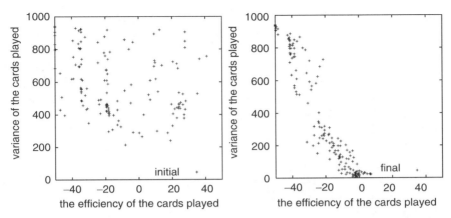

Fig. 2. Individual distributions at the initial (LEFT) and the final (RIGHT) generations

3.2 Fast Evolutionary Programming to Design Appropriate Game Rules

Fast Evolutionary Programming. Fast Evolutionary Programming (FEP) proposed by Xin *et al.* [21] is used in this paper because the Fast Evolutionary Programming is easy to implement and performs well due to the Cauchy distribution mutation. Individuals in the FEP are composed of a pair of real valued vectors (X, η), where X and η indicate the design variables in the problems and variance parameter used in self-adaptive mutation, respectively. (A brief introduction of the FEP is give in the Appendix.)

Utilization of the FEP for Constructing a New Rule of the COMMONS GAME. In this paper, we employ three variables, W_G, A, and W_o to represent new rules. The meaning of them is described as follows:

1. Penalty P_G for green players: We shall propose an appropriate way for penalizing green players which takes the environmental changes into account: $P_G = -W_G \times$ (Gain G), where W_G means the numerical value that is determined by the FEP, and "Gain G" denotes the return that the green players can get if any other player does not choose the black card.
2. Point AOB that black players loose: We shall propose an appropriate way (for asking black players pay cost in trying to penalize green players) which takes the environmental changes into account: $AOB = OB/NOB(OB = -A \times (GainR))$, where A means the numerical value that is determined by the FEP, and NOB means the number of players who have chosen the black cards, and "$GainR$ denotes the return that the red player can get.
3. Point OR that orange players add to the red players: We shall propose an appropriate way (for helping red players maintain the commons) which

takes the environmental changes into account: $OR = W_o \times$ (Gain R), where W_o means the numerical value that is determined by the FEP.

Fitness Evaluation. In order to evaluate the rule parameters mentioned in the previous subsection, 200 games per an individual are carried out. Before runs of the FEP, six neural network players in the final population of MOEA are randomly chosen per game. Namely, 1200 neural network players are selected (including duplicated selection).

After each game, a rule parameter is evaluated by the following:

a) A game in which black cards & orange cards are seldom chosen (almost all of the players only choose a green card or a red card) is quite monotonous. Therefore, the more black (or orange) cards are chosen in a game, the higher value of the evaluation function should be given.
b) A game in which the ranking of each player often changes is exciting. Therefore, the more changes of the top player during a game run, the higher evaluation value should be given. A game in which there is a small difference between the total points of the top player and those of the last player is very exciting. Therefore, the small variance in the total points of each player at the end of a game, the higher the evaluation value should be given.
c) When the environmental deterioration had occurred heavily, each player can receive only a small amount of return. Under such a state of environment, the game should become monotonous. Therefore, a small evaluation value should be given if a game has brought heavy deterioration to the environment. On the other hand, a high evaluation value should be given when a game has brought a moderate final state of the environment.

By taking into account the above points, we have constructed the following evaluation function $T(x)$[5]:

$$T(x) = f(x) + g(x) + h(x) + \alpha(x) + \beta(x),$$

where x denotes a chromosome. The function values of $f(x)$, $g(x)$, $h(x)$, $\alpha(x)$, and $\beta(x)$, correspond to the followings:

$f(x)$: The environmental state at the end of the game;
$g(x)$: The total number of the black card having been chosen;
$h(x)$: The total number of the orange card having been chosen;
$\alpha(x)$: The sum of the variance of the points having been gained by each player;
$\beta(x)$: The total number of the changes of the top player.

3.3 New Rules Obtained by the MOEA and the FEP

In order to find appropriate combinations of the three variables, W_G, A, and W_o, evolutionary search by the FEP has been carried out for 50 generations.

---
[5] In Fig. 3, 5 functions which constitute the evaluation function $T(x)$ are illustrated.

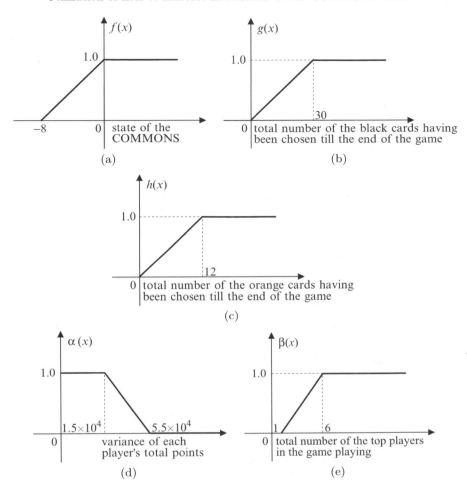

Fig. 3. 5 functions which constitute the evaluation function $T(x)$

The population size of the FEP has been set to be 10. The changes of the average fitness values during the evolution are depicted in Fig. 4. This graph is plotted by averaging over 20 runs. The changes over the whole generations are not so much. However, the main contribution of these changes are caused by the changes of $\beta(x)$. This means that the utilization of the FEP has contributed a lot in changing top players. The reason why other sub-functions, such as $f(x)$, $g(x)$ and so on, did not affect the evolution process is that the neural network players are already sophisticated enough: They play various kinds of cards, including black and orange cards so that the final environmental state has not been deteriorated so seriously.

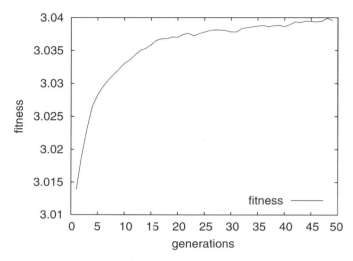

Fig. 4. Changes of averaged fitness value during evolution

We analyzed the best individuals found after the 20 runs. There are two groups in the best individuals: They are around $W_G = 0.27$, $A = 0.04$, and $W_o = 0.12$, and $W_G = 0.29$, $A = 0.2$, and $W_o = 0.1$, respectively. This analysis reveals that much penalization of green players causes the frequent changes of the top players.

3.4 Game Playings Utilizing the New Rules

In order to investigate whether the new rules having been obtained by the MOEA and the FEP are appropriate or not, one of the authors asked his 12 undergraduate students of Osaka Kyoiku University (who had experienced several game playings of the original COMMONS GAME) to play the new games[6] which have been modified by taking the results obtained by the use of the MOEA and the FEP into account. The authors watched the attitudes of the students who participated in the playings of the new games. They felt that almost all of the students concentrated more on the new games than before. After the games, they asked the students for their impressions of the new games.

Answers from the 12 students can be summarized as follows:

(1) 9 students out of 12 expressed their opinions that the new games are far more exciting than the original game.

---

[6] The two new games with the parameter values $W_g = 0.27, A = 0.04$, and $W_o = 0.12$ & $W_g = 0.29, A = 0.2$, and $W_o = 0.1$, respectively, have been played.

(2) Some of the 9 students (who expressed positive impressions toward the new games) explained the reason why they have felt the new games exciting:
  (2-1) In the new games, players can easily choose the black card since cost of using the black card has become rather low. Therefore, all of the players can enjoy dynamic game playing.
  (2-2) In the original COMMONS GAME, penalty point to the green card player by a black card is always -20. However, in the new games, it depends on the environmental state. When the environmental state is 0 or $+1$, damage due to the use of the green card is heavy. This causes the new games rather exciting.
  (2-3) In the new games, points concerning penalty to the green card, cost of using the black card, and etc. are not fixed. Therefore, players should manage to find a good plan for choosing one of the cards in order to adapt successfully to each environmental situation. This causes good thrill to each player.
(3) One of the 3 students (who have shown us negative impressions on the new games) explained the reason why he prefers the original game to the new games: Players (enjoying the game playing of the original game) should be very careful in using black card since its utilization causes them considerable minus point. However, a black card gives heavy minus point to the green card players wherever the state of the environment is. This makes a good thrill to him.
(4) He also added the following reason: In the new games, penalty point to the green card players is comparatively small when the environmental state is negative such as the state "-3" or the state "-4." This causes the situation that players can easily choose the green card.
(5) Many players pointed out the problems involved in the COMMONS GAME: Players who have chosen the red card in almost all of the rounds can easily win the game.

4 Conclusions

In this paper, we have tried to utilize two kinds of EAs, i.e., the MOEA and the FEP for making the COMMONS GAME exciting. The MOEA has been used for generating various types of skilled players. Further, the FEP has been introduced to find out appropriate combinations of the point of each card. As shown in the Fig. 4, we have succeeded in finding highly advanced rules compared with that of the original COMMONS GAME. Several game playings of the COMMONS GAME (using the new rule (derived by using the MOEA and the FEP)) by our students suggest the effectiveness of our approach. However, this has been suggested only by several game playings done by our students. The future research is needed to carry out lots of game playings by various people for the full confirmation of our approach. Further, we should also pay attention carefully to the impressions (concerning the new games) which were posed by our students in order to design a more advanced gaming system.

Acknowledgement

The authors would like to express their thanks to the Foundation for Fusion of Science & Technology (FOST) and the Grant-In-Aid for Scientific Research (C) by Ministry of Education, Science, Sports, and Culture, Japan, who have given them partial financial support.

References

1. Stahl, I., Editor: Operational Gaming: An International Approach, IIASA (1983)
2. Shubik, M.: Games for Society, Business and War, Elsevier (1975)
3. Duke, R.D.: Gaming: The Future's Language, Sage Publications (1974)
4. Hausrath, A.: Venture Simulation in War, Business and Politics, McGraw-Hill (1971)
5. Cohen, K.J. et al.: The Carnegie Tech Management Game, Homewood, Ill.: Richard D. Irwin (1964)
6. Shubik, M.: The Uses and Methods of Gaming. Elsevier (1975)
7. Duke, R.D. and Duke, K.: Development of the Courail Game, Operational Gaming: An International Approach. Stahl, I. Editor, IIASA, (1983) 245–252
8. Powers, R.B., Duss, R.E., and Norton, R.S.: THE COMMONS GAME Manual (1977)
9. Ausubel, J.H.: The Greenhouse Effect: An Educational Board Game. Instruction Booklet, IIASA (1981)
10. Baba, N., Uchida, H., and Sawaragi, Y.: A Gaming Approach to the Acid Rain Problem. Simulation & Games **15** (1984) 305–314
11. Baba, N.: The commons game by microcomputer. Simulation & Games, **15** (1984) 487–492
12. Baba, N. et al.: Two Micro-Computer Based Games, Working Paper, WP-86-79, IIASA (1986)
13. Fogel, D.B.: Evolving behaviors in the iterated prisoner's dilemma. Evolutionary Computation, **1(1)** (1993) 77–97
14. Chong, S.Y., Yao, X.: Behavioral diversity, choices and noise in the iterated prisoner's dilemma. IEEE Trans. Evolutionary Computation **9(5)** (2005) 540–551
15. Barone, L., While, L.: Adaptive learning for poker. Proceedings of the Genetic and Evolutionary Computation Conference (2000) 560–573
16. Moriarty, D. and Miikkulainen, R.: Discovering complex othello strategies through evolutionary neural networks. Connection Science **7(3)** (1995) 195–209
17. Lucas, S.M., Kendall, G.: Evolutionary computation and games. IEEE Computational Intelligence Magazine **1(1)** (2006) 10–18

18. Deb, K., Agrawal, S., Pratap, A., and Meyarivan, T.: A Fast and Elitist multi-objective Genetic Algorithm: NSGA-II. IEEE Trans. Evolutionary Computation **6(2)** (2002) 182–197
19. Coello Coello, C.A.: Evolutionary multi-objective optimization: a historical view of the field. IEEE Computational Intelligence Magazine **1(1)** (2006) 28–36
20. Bäck, T., and Schwefel, H.-P.: An overview of evolutionary algorithms for parameter optimization. Evolutionary Computation **1(1)** (1993) 1–23
21. Yao, X., Liu, Y., and Lin, G.: Evolutionary programming made faster. IEEE Trans. Evolutionary Computation **3(2)** (1999) 82–102
22. Iida, H., Takeshita, N., and Yoshimura, J.: A Metric for Entertainment of Boardgames: Its Implication for Evolution of Chess Variants. Entertainment Computing: Technologies and Applications, pp. 65-72 (2002)
23. Rauterberg M.: About a framework for information and information processing of learning systems. E. Falkenberg, W. Hesse, A. Olive (eds.), Information System Concepts - Towards a consolidation of views, pp. 54-69 (1995)
24. Tesauro, G.: Temporal Difference Learning and TD-Gammon. Communications of The ACM, **38** (1995) 58p-68
25. Baba, N.: An Application of Artificial Neural Network to Gaming. Proc. SPIE Conference, Invited Paper, **(2492)** (1995) 465–476
26. Baba, N., Kita, T., and Takagawara, Y.: Computer Simulation Gaming Systems Utilizing Neural Networks & Genetic Algorithms. Proc. SPIE Conference, Invited Paper, **2760** (1996) 495–505
27. Baba, N.: Application of Neural Networks to Computer Gaming, Soft Computing in Systems and Control Technology, Tzafestas S.G., Editor, World Scientific Publishing Company, Chapter 13 (1999) 379–396
28. Baba, N. and Jain, L.C., Editors: Computational Intelligence in Games, Springer-Verlag (2001)
29. Baba, N. and Shuqin, W.: Utilization of Neural Networks and Genetic Algorithms to Make Game Playing Much More Exciting, Gaming, Agent Based Modeling & Methodology, Arai, K. *et al*, Editors, Springer-Verlag, Chapter 22 (2005) 207–216

Appendix: Brief Introductions Concerning the NSGA-II and the FEP

(A.1) The Nondominated Sorting Genetic Algorithm II (The NSGA-II)

Below, we briefly introduce the NSGA-II (Nondominated Sorting Genetic Algorithm II) proposed by Deb et al. [18]. The NSGA-II is one of the Multi-Objective Evolutionary Algorithms. It is characterized by selection mechanism: In order to select a new population for preserving diversity among solutions, two indices for the individual i are utilized: nondomination rank i_{rank} and crowding distance $i_{distance}$. The nondomination rank i_{rank} of an individual i indicates the number "1 + the number of individuals which dominate the individual i." For instance, the nondomination rank i_{rank} of the individual i in the Pareto set in the combined population is 1. Individuals whose nondomination rank is 2 are dominated only by an individual in the Pareto set. The crowding distance $i_{distance}$ denotes the average distance of the two individuals on either side of the individual i along each of the objectives.

In the selection mechanism of NSGA-II, a partial order $\prec_n$, called Crowded-Comparison Operator, is introduced by using the two indices i_{rank} and $i_{distance}$.

$i \prec_n j$ if $(i_{rank} < j_{rank})$ or $((i_{rank} = j_{rank})$ and $(i_{distance} > j_{distance}))$

In order to calculate the above indices and utilize them efficiently for the selection mechanism, two sorting algorithms are proposed by Deb et al. For those interested in further details, we recommend reading the paper written by Deb et al. [18].

(A.2) Fast Evolutionary Programming (The FEP)

Below, we show the algorithm of the FEP [21].

1. Generate the initial population consisting of μ individuals.
2. Evaluate each individual.
3. Let each individual (X, η) create an offspring (X', η') as:

$$x'_j = x_j + \eta_j \delta_j \qquad (A.2.1)$$
$$\eta'_j = \eta_j \exp(\tau' N(0,1) + \tau N_j(0,1)) \qquad (A.2.2)$$

where x_j and η_j denote the jth component of vectors X and η, respectively. δ_j, $N(0,1)$, and $N_j(0,1)$ denote a Cauchy random variable, a standard Gaussian random variable, the jth independent identically distributed standard Gaussian random variable, respectively. In (A.2.2), coefficients τ and τ' are set to be $(\sqrt{2\sqrt{n}})^{-1}$ and $(\sqrt{2n})^{-1}$, respectively.
4. Evaluate each offspring.

5. Conduct pairwise comparison over all of the parents and offsprings. In order to evaluate each individual, q opponents are chosen randomly. For each comparison, each individual receives "win" when its fitness value is higher than that of the opponent.
6. Pick up μ individuals from the set of the parents and the offsprings by taking the ranking due to the number of the winnings into account.
7. Stop if halting condition is satisfied. Otherwise go to Step 3.

Adaptive Generation of Dilemma-based Interactive Narratives

Heather Barber and Daniel Kudenko

University of York, Heslington, York, YO10 5DD, England
{hmbarber, kudenko}@cs.york.ac.uk

Abstract. In this paper we present a system which automatically generates interactive stories that are focused around dilemmas to create dramatic tension. A story designer provides the background of the story world, such as information on characters and their relations, objects, and actions. In addition, our system is provided with knowledge of generic story actions and dilemmas which are currently based on those clichés encountered in many of today's soap operas. These dilemmas and story actions are instantiated for the given story world and a story planner creates sequences of actions that lead to dilemmas. The user interacts with the story by making decisions on these dilemmas. Using this input, the system adapts future story lines according to the user's preferences.

1 Introduction

In this paper we introduce the concept of interactive drama and discuss the shortcomings of prominent previous work [1, 3, 4, 5, 7, 8, 9, 10, 11, 12, 13, 14]). We then propose a system which overcomes these shortcomings to generate interactive stories which are long (potentially infinitely so), and that adapt to the user's behaviour. To add dramatic tension, the story incorporates dilemmas as decision points for the user. These dilemmas can be based on the clichés found in many contemporary soap operas, such as the trade off between personal gain and loyalty to a friend. Overarcing stories connect these dilemmas as points of interaction within a coherent plotline that is dynamically created, based on the user's response.

Our goal is to keep the story designer's input to a minimum. In the proposed system, the story designer provides the story background in the form of character information and other knowledge that relates to the world in which the story is to be created (e.g., a pirate ship). The system then instantiates all generic knowledge on story actions and dilemmas accordingly and thus creates the narrative.

Our paper is structured as follows. We first give an introduction to the motivations for and the definition of interactive drama. This is followed by an overview of previous research in the area, and the shortcomings thereof. A general overview of our system is then given, followed by a discussion on how the story background is specified. We then proceed with a description of dilemmas, the story generator, and the user modelling component. We finish the paper with conclusions and a brief discussion of future work.

2 Interactive Drama

2.1 Stories

Storytelling is appreciated by many people, as both teller and audience. Before writing, stories were told orally – with audience participation. It is still true that listening to a friend narrating a story is an enjoyable way to spend time. As storytelling has evolved – through drama, writing, print, film and television – interactivity has been neglected.

Receivers (listeners, readers or viewers) of a story will often believe that they know best as to how the story characters should act, and can be frustrated by their lack of control over this. They may want to become more involved in the storyworld, to play a more active part – perhaps even to become a character. An interactive story offers a world in which the participant can have a real effect – both long and short term – on the story which they are experiencing. It is often said that the best stories transport their receiver into the world of the story, yet the actions of the characters may be contrary to what the receiver would imagine. Interactive drama overcomes this problem and offers a unique, exciting and participatory story.

2.2 Computer Games

Most modern computer games involve a story. In general this is an essentially linear story or series of stories. As a result, this element generally violates a basic requirement for such games – the need for interaction and a clear effect of such interactions. A player who is experiencing a game with a story element will expect to have a genuine impact on that story. Otherwise the player's enjoyment of the game, and their agency within the game world, will be reduced. There are games with no explicit story structure – where the player is encouraged to perceive their own stories within the world. These stories are truly interactive but lack the skill of a playwright and subsequent high level of dramatic interest. An interactive drama would combine the free interactions of the player with this play writing skill to create a dramatically interesting game playing experience.

2.3 Interactive Drama

An interactive drama is a game world in which the user-controlled character(s) can physically and mentally interact with (perceived) total freedom while experiencing a dramatically interesting story which is fundamentally different on every play – dependent on the user's actions.

An interactive drama is a game world..., it will be set within a computer-simulated virtual world. The exact depiction of this world will depend on the genre of story to be experienced – which it would ideally be possible for the user to select.

...in which the user-controlled character(s)... It may be possible for the user to control more than one character and to change which they control. A higher-level scenario in which the user will control groups rather than a specific character is possible but is less likely to result in a dramatically interesting story as most stories centralise on characters. It is also probable that the user will be more emotionally involved in a game world in which they participate at a character level.

...can physically and mentally interact..., ideally being able to speak freely and be understood within the game world. Characters will be able to move around the world as and when they desire. Interactions with other game world characters are essential as stories rarely involve a single character.

...with (perceived) total freedom... The user of an interactive drama is likely to become frustrated if they are not free to act as and when they wish. It is not entirely true that they require total freedom within the game world , as what is important is the user perception of freedom. Laurel [12] explains that it 'is difficult to imagine life, even a fantasy life, in the absence of any constraints at all', giving the example of gravity within a game world as being generally taken for granted rather than seen as limiting a user's freedom. There may be other implicit constraints on a user, but as long as these are in keeping with the game world and the user's perception thereof the user will still believe that they are free within the world.

...while experiencing a dramatically interesting story..., as a boring story – while potentially an interactive drama – is not considered to be sufficient here. For the system to be successful the story experienced by the user must be dramatically interesting for that user. Despite a wide range of research there are no absolute rules for the creation of a dramatically interesting story but rather various guidelines that can be followed.

...which is fundamentally different on every play..., due to the clear and genuine effect of the user's actions in the game world. Current computer games tend to present a story which remains the same each time. There may be various options within the story – at various stages or even in the potential for a range of endings. At a low level, it is likely that the user will have a different experience each time, particularly as they experiment with the world to see what is possible. However, the aim in an interactive drama is not to have the same story with a few variations but to have a story which is entirely different each time, to the extent that it exists as a separate story in its own right.

...– dependent on the user's actions. It is not sufficient for the story to be different simply because the system is capable of creating original stories.

These stories must be different dependent on the user's actions. The user must be exploring a world in which an infinite range of stories can take place (which are not at all pre-defined) and where which story takes place will genuinely depend on the way the user acts within that world. The user will be able to act as and when they desire in ways which will have a perceivable long and short term effect on the story.

3 Related Work

There are four fundamental issues which should be addressed in the creation of an interactive narrative. It is important for an interactive narrative to adapt to the user so that the experience is dramatically interesting for them personally. A truly original story must be created so that the user will be able to re-play the system and have a uniquely enjoyable experience each time. The story's dramatic interest should be sustainable over a longer time period, as in conventional media. This will give the user the chance to become more involved in the storyworld. In order to appeal to a wide audience the basic techniques used in the system should be easily transferable to further domains and storyworlds – this is also in keeping with conventional storytelling techniques.

Although some existing interactive drama systems have addressed one or two of these issues none of the techniques used in their creation are capable of addressing all of these failures. In particular, scalability to a longer term story is not possible with any of the systems and transferability is severely limited. Subsections 3.1 to 3.4 give more detail on each issue.

3.1 User Adaption

A user model can be employed as a method of creating a story which is of greater dramatic interest to the user. This means that the story will be specifically created for the user and their individual preferences.

In the IDA [8] system the user becomes a ghost and must solve a murder mystery. The story structure is provided by a plot graph. A user model is employed to predict which moves and choices the user is most likely to make and thus to direct them towards a particular storyline. So, for example, if a user has never been into the library but some event is required to take place there, perhaps music could be played in the library. These techniques will become less subtle as the story moves closer to the time when the events are required (by the plot graph) to take place and ideally the user will be modelled well enough for these more blatant techniques to be unnecessary.

In IDtension [12] the user selects actions (for any characters) alternately with the system in attempting to achieve the goals of the story. This system employs a user model in order to increase the dramatic tension experienced by the user. If the user, for instance, constantly avoids actions which involve violence then the system will present them with a conflict in which they must be violent in order to achieve the goals of the story. Such conflicts provide the drama in the story and it is thus a great step forward from conventional literature to have these specifically tailored to the user and their preferences.

The stories experienced by users of these two systems do to some extent adapt to the user, their expected actions and perceived preferences. None of the techniques used by other systems have the potential to adapt their stories to the individual user.

3.2 Truly Original Story

Stories are to some extent re-experiencable. A good book will be read again and again, and this repeatability is also true of certain plays and films. A good computer game with a reasonably strong story element is not generally re-playable. This is because it will quickly become apparent how limited the player really was in their experience, and how little genuine freedom they have within the game world. Board games and computer games without a story element (for example solitaire or Tetris) will be played again and again. The ideal interactive drama will have this element of re-playability. It will present the user will an entirely new and unique story on each experience.

The Erasmatron [4] and DEFACTO [11] systems produce a dynamically generated story, which is thus original on each experience. The DEFACTO system is used to graphically present stories to the user after they have selected all of their actions, with a twist which would quickly become predictable. The OPIATE [5] stories are also dynamically generated, according to a general pattern for the fairy tale genre. The remaining systems use a more fixed plot structure and the fundamental basis of the system is thus contradictory to the production of an original story.

3.3 Scalability

Traditional stories vary in length depending on the medium. A book could take days to read, a film lasts 2-3 hours, and soap operas continue infinitely. There are short versions but for true richness of storyworlds the longer is necessary. Interactive dramas created thus far only last a few minutes at the most. The Façade system involves the user in a couple's marital issues. The stories produced are by far the longest but massive amounts of work are required for each few seconds. All of the methods used in existing systems would require too much pre-definition to be capable of sustaining a dramatically interesting story over a longer time period.

3.4 Transferability

Ideally the techniques used in an interactive drama would be applicable to any story genre. This would mean that the stories could be enjoyed by a much greater range of users and that the possibilities would be more extensive. The techniques used in creating stories in traditional mediums can usually be applied to any genre, for example films can be: horror, fantasy, romance, comedy, adventure, tragedy, etc.

The mimesis [14] system is designed as a general architecture able to interact with any game engine. The Erasmatron [4] system has been applied to a range of stories but requires a large amount of author pre-definition for this, meaning that there is extensive author involvement required before a storyworld can be created. It would be a vast improvement if the system was capable of creating this content.

For the majority of the systems, applying the techniques to additional domains would require rewriting almost the entirety of the system. This has been demonstrated by the I-storytelling [3] and Façade [9] groups' creation of new storyworlds – with very little reusability from previous work.

3.5 Planning

The system introduced in this paper utilises planning techniques in the creation of an interactive narrative. Other interactive drama systems in existence use planning. Mimesis [14] uses planning to achieve the story goals. This is much longer-term planning and is less flexible around the user's interactions - which will either be accommodated in re-planning or intervened with. In the I-Storytelling [3] system, hierarchical task network (HTN) planning is used. Each character is equipped with an HTN to follow in the story, which is defined before the story begins. There is very little allowance for user interactions in this system. In neither system is there any allowance for the story to be dynamically created, but only for it to be dynamically adjusted.

More recent systems use planning techniques to create stories in collaboration with a user. In [13] the planner is used to create each stage of a planning graph. The user is then able to choose from the subsequent options to decide which will appear in the final version of the story. The story presentation will be a mimesis-style experience. Points for re-planning and intervention by the system are specified by the user at the story creation stage, whereever a need is identified by the system. The shortcomings of Mimesis apply here also. The system described in [7] involves goal events which are planned for. The user is able to specify some of these events and to prompt re-planning for any. They may be ignored. The user must then select the final ordering of events - given any constraints. The resulting story is then graphically produced without any interaction, and at a much lower level than that at which the user aided in the story creation.

Fairclough's system [5] utilises planning techniques to dynamically create an interactive story in the fairy tale genre. There are a finite number of subplots and the user's actions determine which is experienced. A plan is then created for the subplot, which consists of a "sequence of character actions" given to the characters (other than the user) as goals. The user has a high level of freedom but they are not entirely flexible as they must adhere to a limited number of subplots. In contrast, the system proposed in this paper will allow the user complete freedom. The user is also modelled so that the experience is more enjoyable for them personally. The dilemmas posed to the user in our system will increase the dramatic interest of the stories.

4 System Overview

The interactive drama knowledge base consists of: the storyworld (which contains information regarding the characters); story actions; and dilemmas which can occur in the storyworld. This information is partially genre dependent and provided by the story designer, with the remainder being hard coded. These components are all drawn upon in the generation of a narrative through planning. The user is able to interact with the narrative generator, and their actions effect the story experienced. A user model is employed to ensure that the story's dramatic interest is maximised. The interactions between the system components are shown in figure 1. Each of these components is discussed further in the following sections.

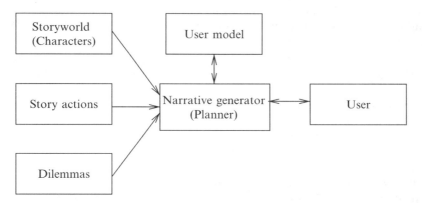

Fig. 1. This figure shows the components of the system and how they interact

5 The Storyworld

The storyworld consists of characters and locations at which the characters can be. These characters have various associated traits, as detailed here.

- Each character's associated domain independent attributes can include information such as attractiveness, gender, sexuality and age group.
- Characteristics are slightly more variable, for example: generosity, morality and selfishness.

- It is possible to specify particular personalities, such as bad_boy and busybody. These genre specific character descriptions are those which are not fully deducible from other character traits and which will relate to specific storylines within the current domain.
- Characters have storyworld relationships with one another, including friendship and love. They are able to disapprove of one another's partnerships. This can be for any one of a variety of reasons, including an age difference or snobbery. Relationships are unidirectional and have an associated strength, although feelings of one character for another will affect the reciprocity.
- The characters hold storyworld principles, such as monogamy. Under specified pressures and circumstances, principles can be broken (or their associated strength of belief reduced). Characters also have aspirations, for example wanting a baby. These affect which actions a character will participate in and the dilemmas in which they will become involved.

A range of values is associated with each attribute and characteristic. A character's nature affects which actions they can participate in and also, ideally, the user's opinion of that character. The character's personal traits should be apparent to the user from the way the character acts within the storyworld. Each character should act in a manner which is consistent with their traits and how they have acted previously, while at the same time avoiding predictability.

A series of genre-specific locations are required by the storyworld. At any given time in the story, each character will be at one of these locations. Direct interactions between characters can only take place if they are at the same location.

6 Actions

Those actions which can take place within the storyworld must be specified for each domain. Every possible action should be included and although these will vary between domains there will be a significant overlap.

The domain specific storyworld actions can include characters falling in love, becoming pregnant and being involved in crimes – such as drugging or murder. Each of these actions has associated conditions which must be satisfied before execution (preconditions) and effects which represent changes to the storyworld following execution. For example, the action of a character moving between locations l and k has preconditions of the character being at location l and there existing a path between locations l and k. The effects of this action are that the character is at location k and is no longer at location l. This follows the STRIPS representation.

Before an action is made available to the system for use within a storyline an applicability check is carried out. This ensures that the action is of the type that the acting character is likely to make. For example, a more attractive character may start to fancy a very generous character. A character's attributes, characteristics and personalities will affect which actions are possible for that particular character as an action can only be utilised if its applicability is high enough for that character.

7 Dilemmas

Field [6] states that "drama is conflict", that the dramatic interest in a story centralises on its conflicts. In genres which make use of clichéd storylines these are usually found to be essentially conflicts (or dilemmas). Writers utilise these dilemmas in the creation of stories. A general form of each such clichéd dilemma can be determined, and a computerised storywriter can create an interactive drama around these.

Since the focal point of an interactive drama is the user, each dilemma should represent a conflict to that user. Within the course of the experience, they will be required to make fundamentally difficult decisions which will have negative outcomes whatever choice they make. There may also be decisions in which the user has to decide how to distribute limited benefits in different areas or to different characters.

Our experience showed that when more than two characters were involved in a dilemma, it was either expandable to multiple two character dilemmas, or the characters receiving payoffs naturally divided into two groups with the same resultant utility. Therefore a user decision on a dilemma will involve only two recipients of utility payoffs. Five such dilemma categories were identified. These do not consist of all payoff matrices for two users, as many such matrices would not involve a dilemma for the character making the decision. The relevant categories are: Betrayal (dilemma 1), Sacrifice (dilemma 2), Greater_Good (dilemma 3), Take_Down (dilemma 4) and Favour (dilemma 5).

In these dilemmas: A_X represents the decision of character X being to take action A; u_C^i represents the utility of character C for the respective action; and i denotes the relative value of the utility, i.e., u_C^1 is greater than u_C^2.

Betrayal

When presented with a Betrayal dilemma a character X must decide whether or not to take an action A_X which would result in their best possible utility while considering that this would simultaneously be the worst possible outcome for their friend (or someone close to them) Y. The dilemma would clearly not exist were X and Y not friends. This dilemma can be represented as a payoff matrix as shown in equation 1.

$$\begin{array}{c|c} A_X & (u_X^1, u_Y^2) \\ \hline \neg A_X & (u_X^2, u_Y^1) \end{array} \land friends(X, Y) \qquad (1)$$

A character having the opportunity to be unfaithful to their partner is an example of the Betrayal dilemma.

Sacrifice

A character X facing the Sacrifice dilemma is able to choose an action A_X which will result in their worst possible utility but also the best possible outcome for their friend Y. It is a necessary condition of this being a dilemma that these characters are friends. The payoff matrix for this dilemma is shown in equation 2.

$$\begin{array}{c|c} A_X & (u_X^2, u_Y^1) \\ \hline \neg A_X & (u_X^1, u_Y^2) \end{array} \wedge friends(X, Y) \qquad (2)$$

An example of the Sacrifice dilemma occurs when a character has committed a crime which their friend has been accused of. Here a character has the opportunity to admit to their crime and thus accept the punishment rather than allowing their friend to take the blame.

Greater Good

Involvement in a Greater Good dilemma means that a character X is able to take an action A_X which will result in their best possible utility but also the best possible outcome for their enemy, Y. This dilemma would obviously not exist if the characters were not enemies. It is represented as a payoff matrix in equation 3.

$$\begin{array}{c|c} A_X & (u_X^1, u_Y^1) \\ \hline \neg A_X & (u_X^2, u_Y^2) \end{array} \wedge enemies(X, Y) \qquad (3)$$

A character deciding whether to give something (such as information or a friend) to their enemy Y in order to save themselves (and possibly also their family) would be experiencing the Greater Good dilemma.

Take Down

In the Take Down dilemma a character X has the option of taking an action A_X which will result in their worst possible utility but also the worst outcome for their enemy Y. Clearly the characters must be enemies for the dilemma to exist. The payoff matrix for this dilemma is given in equation 4.

$$\begin{array}{c|c} A_X & (u_X^2, u_Y^2) \\ \hline \neg A_X & (u_X^1, u_Y^1) \end{array} \wedge enemies(X, Y) \qquad (4)$$

A character deciding whether to injure (or even kill) their enemy in full awareness that they will receive a punishment for this crime would be involved in the Take Down dilemma.

Favour

The favour dilemma sees a character X able to choose whether or not to take an action A_X. This character will not receive any direct utility from their action regardless of their choice, but the utilities of characters Y and Z will change as a result of this action decision. If character X chooses to take the action A_X the utility of character Y will be Y's highest possible and character Z will receive their lowest utility - and vice versa if X chooses not to take this action. Equation 5 gives the payoff matrix for this dilemma.

$$\begin{array}{c|c} A_X & (u_Y^1, u_Z^2) \\ \hline \neg A_X & (u_Y^2, u_Z^1) \end{array} \quad (5)$$

An instance of this dilemma occurs when a character must choose between potential partners. It is necessary that there is no discernible benefit to the character making the decision of choosing one partner over the other.

As can be seen, dilemmas 1 and 2 are the inverse of one another, as are dilemmas 3 and 4. This means that any dilemma which falls into one of these categories can be inverted to become a dilemma of the other category. All five categories are kept to increase ease of dilemma identification within specific genres. From these categories (as given in equations 1 to 5) dilemma instances can be found and generalised within each domain. From the generalised form of the dilemma the system will be able to create new dilemmas. In the presentation of these to the user wholly original stories are created.

It will not be possible to create great literature in this way – the use of clichéd storylines prevents this. However, such stories are enjoyed by many people and this method is common in such genres as James Bond films, soap operas (soaps) and "chick flicks". The story is built around the cliché, and it is the cliché as well as the story which the audience appreciate, the very repetitiveness and familiarity of the dilemmas adding to the dramatic interest. It can only be imagined how much more enjoyment could arise from the user becoming a character in such domains, and experiencing the dilemmas first hand.

8 The Narrative Generator

Prior to a dilemma being presented to the user certain conditions must be met within the storyworld. These are the preconditions of the dilemma. It is the task of the storywriting system to achieve these preconditions. This constitutes the build-up – the essence of the story itself. Given actions within the storyworld, the system can use planning to satisfy a dilemma's preconditions. In this way, a plan to achieve a dilemma becomes a storyline. The interactive drama is made up of a series of such substories, dynamically selected according to dramatic interest.

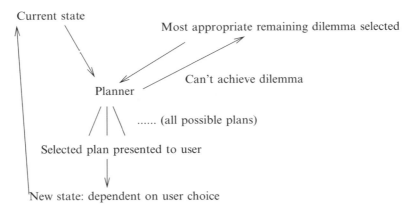

Fig. 2. This figure gives an overview of the system moving between states dependent on plans, dilemmas and user decisions

The system uses a GraphPlan planner [2] which utilises a STRIPS-style representation of actions. The GraphPlan algorithm is sound and complete. On being passed a dilemma, the planner finds all plans to achieve this dilemma given the current storyworld state and background knowledge. Dilemmas are not running concurrently, therefore the build-up cannot be too extensive as the user could become disinterested in the experience. As a result, the system is designed to prefer shorter plans. Stories of the same length will involve more drama if plotlines are shorter. From these plans, that which is most dramatically interesting can be selected. Once the user has made their choice, the system updates the storyworld state in accordance with that choice. The system can then plan from the new state in order to be able to present another dilemma to the user – thus continuing the interactive drama. This sequence of events is demonstrated in fig. 2. From this figure it can be seen that the planner finds all plans in the story dependent on the current state and a given dilemma. If no plan can be found for this dilemma, another will be selected. Once all plans have been found, the most dramatically interesting can be presented to the user, resulting in a new state from which the story will continue.

The potential consequences of each decision must be clear to the user before they make their choice. Once they have chosen, these repercussions on the storyworld will be implemented. The resultant state is thus entirely dependent on the user's decision.

The sequence in which the dilemmas are selected for planning is dependent on the story history, the frequency of dilemma use, dramatic interest and the user model. Dilemmas must depend on what has happened previously and thus become part of a consistent story. Certain dilemmas will only occur occasionally, others will be more frequent. This will need to be determined for each domain, and considered when selecting each dilemma. It is necessary to plan for dilemmas which have a high dramatic interest, although this will largely depend on the current user. The user model is discussed in the next section.

9 The User Model

The user of an interactive drama system should be modelled rather than controlled. The story should adapt to the user's interactions rather than forcing the user to follow a particular storyline.

The user model will ideally be used to identify which dilemmas are going to be most conflicting and dramatically interesting for the current user. There will be an "interestingness" value associated with each dilemma. This value is initially fixed but will adapt to suit the user and their modelled personality. The system will search for the most interesting story path to a pre-defined fixed depth (dependent on the size of the search space and the speed of the search and planning algorithms).

Each dilemma has associated assumptions as to how the modelled values will change dependent on the user decision. Once they have made their choice, the user model will be updated accordingly. A selection probability will be associated with each criterion, so that the credibility given to the user model will depend on how many times it has been updated. It may additionally depend on how recently the criterion being utilised was updated – since the user and their opinions may well change through the course of the interactive drama. This user model will then be employed to approximate the probability of a user making a particular choice within a dilemma. It then calculates the expected total "interestingness" of that path. The system will select that dilemma which has the highest chance of leading to the most dramatically interesting experience for the user. A section of this search is shown graphically in fig. 3.

Those aspects of the user which are modelled must be general enough to apply to a number of dilemmas, but also specific enough to accurately predict which choices the user is likely to make. Personality traits included in this general model include: honesty, responsibility_for_actions, faithfulness, strength_of_principle, selfishness, preference_for_relationship_or_friendship, strength_of_character and morality. There may also be domain specific factors. In addition, the user's opinions of other characters must be modelled, for example: value_for_relationship, value_of_friendship, value_of_love. Each of these criteria will have an associated value, which for certain criteria will take variables, for instance: strength_of_principle will take a separate value for each principle; value_of_friendship and value_of_love will take a separate value for each character within the storyworld.

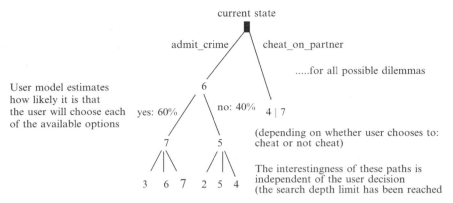

Fig. 3. This figure shows a section of a potential user model. The expected interestingness of each dilemma is as given. The highest prospected score of the admit_crime dilemma is 12.2. A similar calculation can be carried out for each path, and the most interesting subsequently selected

10 Example

The techniques discussed here are applicable in any genre which places a particular emphasis on stereotypes and clichés. It was decided to initially focus on the creation of an interactive soap. This domain does not require an overall story arc but rather involves an infinite series of 'mini-stories'.

The domain of soap operas is commonly understood to revolve around stereotypical storylines. In many cases, these will involve a character being presented with a decision likely to result in negative outcomes either way. A range of such dilemmas from Neighbours, Home and Away, Coronation Street, Eastenders and Hollyoaks have been identified and generalised. These soaps were selected for their accessibility, familiarity and popularity with the general public.

For each soap, a series of dilemmas which characters had faced in recent years were identified[1]. It was found that these dilemmas fell into only three of the five possible categories, namely Betrayal (1), Sacrifice (2) and Favour (5). Figure 4 gives examples of these dilemmas, one of which is generalised in fig. 5.

All domain specific background knowledge was added to the system, including STRIPS-style actions (such as why two characters may fall in love) and locations (for example club and house) which appear in the considered soaps. In fig. 6 a Prolog action from the system is shown with its pre- and postconditions.

---

[1] Thanks to George Barber for his knowledge of soaps and ability to identify such dilemmas.

Hollyoaks: Becca has the opportunity to cheat on her husband Jake with Justin, a schoolboy in her class.
Eastenders: Jane has to decide whether or not to cheat on her husband Ian with the local bad boy Grant.
Coronation Street: Danny has the opportunity to cheat on his wife with Leanne, his son's girlfriend.
Home and Away: Kim has to decide whether or not to cheat on his girlfriend with his best friend Hayley.
Neighbours: Stu has the opportunity to cheat on his institutionalised wife Cindy with a local pretty girl – who previously went out with his brother.

Fig. 4. As can be seen from this small sample of similar dilemmas, the plotline of a character being presented with a dilemma involving cheating on their partner has been used in all of the examined soaps. This demonstrates the frequent use of clichéd storylines in soaps

```
A_X : cheat_on_partner(character(X))
preconditions: partners(X,Y) ∧ loves(X,Z) ∧ loves(Z,X)
dilemma: ''Would you like to cheat on your partner character(Y) with
character(Z) who loves you?''
if user chooses to cheat:
add to state: cheating(X,Y,Z)
update user model:
honesty - lowered, faithfulness - lowered, value_for_relationship with
Y - lowered
if user chooses not to cheat:
delete from state: loves(X,Z)
update user model:
honesty - raised, faithfulness - raised, value_for_relationship with Y
- raised
```

Fig. 5. A dilemma of type Betrayal which is frequently used in soaps (see fig. 1), and can be presented to the user of this interactive drama system

```
Action: X starts to fancy Y
Preconds: fancies(Y,X) ∧ attractiveness X > 1 ∧ attractiveness Y = 1
Effects: fancies(X,Y)
```

Fig. 6. An action in the STRIPS representation in which any characters in the system can participate. Here, an attractive person is in love with someone less attractive. In a soap world (where looks are very important) the less attractive character will fall in reciprocal love with the more attractive

Once a plan for a dilemma has been found, the system will execute this dilemma. In order to do so, the current state must be updated in accordance with each action in the plan. These actions are shown to the user, so that they understand the dilemma and will consequently be more interested in it.

(a)
Background knowledge (includes): character(adam), attractive(adam,0),
character(joe)
Current state (includes): partners(joe,eve)
User model (includes): character(eve), attractive(eve,1),
loves(eve,adam)
(b)
Action: fall_in_love(adam,eve)
You are going out with joe. But you are also in love with adam who
loves you. Will you cheat on your partner joe with adam?

Fig. 7. The necessary background knowledge and state elements are given here (a), prior to the build-up to and presentation of a user dilemma (b)

Figure 7 shows the build-up to and presentation of a sample dilemma to the user. Appendices A and B give the story experienced by one user of the system.

11 Conclusions

In this paper, we presented an interactive narrative generator that is able to create long, and potentially infinite, story lines that incorporate dilemmas to add dramatic tension. The stories are dynamically created based on user decisions, and adapt to the user's tendencies.

In future work we plan to increase the interactivity of the system, giving the user the opportunity to influence the story between dilemmas with their actions. There is also the potential for the creation of soap-specific dramas, with characters as in real soaps, for example an interactive *Eastenders* soap. The user model will be extended, to assess interestingness value of dilemmas to the individual user.

It is ultimately intended that these interactive drama worlds will be graphically simulated. In this way the user will see the storyworld as in conventional media but will be a character, and will be able to act as such. In the short term basic text descriptions will be used, which may be translated into their pictorial representation.

References

1. Joseph Bates. Virtual reality, art, and entertainment. *Presence: The Journal of Teleoperators and Virtual Environments*, 1, 1992.
2. A. Blum and M. Furst. Fast planning through planning graph analysis. *Artificial Intelligence*, 90, 1997.
3. Marc Cavazza and Fred Charles. Character-based interactive storytelling. *IEEE Intelligent Systems*, 17, 2002.
4. Chris Crawford. *Chris Crawford on Interactive Storytelling*. New Riders, 2004.

5. Chris Fairclough. *Story Games and the OPIATE System*. PhD thesis, University of Dublin - Trinity College, 2004.
6. Syd Field. *The Screen-writer's Workbook*. Dell Publishing, New York, 1984.
7. Börje Karlsson, Angelo E. M. Ciarlini, Bruno Feijó, and Antonio L. Furtado. Applying a plan-recognition/plan-generation paradigm to interactive storytelling. In *Workshop on AI Planning for Computer Games and Synthetic Characters*, The Lake District, UK, 2006.
8. Brian Magerko. Story representation and interactive drama. In *1st Artificial Intelligence and Interactive Digital Entertainment Conference*, Marina Del Rey, California, 2005.
9. Michael Mateas and Andrew Stern. Façade: An experiment in building a fully-realized interactive drama. *Game Developers Conference, Game Design track*, 2003.
10. D. Rousseau and Barbara Hayes-Roth. A social-psychological model for synthetic actors. In *International Conference on Autonomous Agents*, St. Paul, Minnepolis, 1998.
11. Nikitas M. Sgouros. Dynamic, user-centered resolution in interactive stories. *International Joint Conference on Artificial Intelligence*, 2, 1997.
12. Nicolas Szilas. Idtension: a narrative engine for interactive drama. In *1st International Conference on Technologies for Interactive Digital Storytelling and Entertainment*, Darmstadt, Germany, 2003.
13. James M. Thomas and R. Michael Young. Author in the loop: Using mixed-initiative planning to improve interactive narrative. In *Workshop on AI Planning for Computer Games and Synthetic Characters*, The Lake District, UK, 2006.
14. R. Michael Young. An overview of the mimesis architecture: Integrating intelligent narrative control into an existing gaming environment. In *The Working Notes of the AAAI Spring Symposium on Artificial Intelligence and Interactive Entertainment*, Stanford, California, 2001.

A The Storyworld State

The state of the storyworld at any given time is given in this easily extendible form. Each character is shown with their location, feelings towards others and the principles they currently hold. For example: John is in the park; likes Joe and you (the user); fancies you; and holds the principle of not stealing. Ultimately the user's principles and feelings for other characters will be modelled rather than explicitly stated.

```
Adam is at location house
Adam likes: Joe and John
Adam fancies: Joe and John
Adam does not hold the principle of not stealing

Joe is at location shop
Joe likes: Adam, Fred, John and you
```

```
Joe fancies: Fred
Joe does not hold the principle of not stealing

Fred is at location shed
Fred likes: Adam and you
Fred fancies: Adam, Joe and John
Fred does not hold the principle of not stealing

John is at location park
John likes: Joe and you
John fancies: You
John holds the principle of not stealing

You are at location jail
You like: Joe
You fancy: Joe and John
You hold the principle of not stealing
```

B A Sample Story

An extract from an interactive narrative experienced by Eve as produced by the system is given here. For clarity and brevity the state of the storyworld is not given, although it does begin from the state shown in appendix A. User input is given in italics.

```
Action is Adam moves between house and club
Action is Eve moves between jail and club
Action is Fred moves between shed and club
Action is Adam gets drunk
Action is Eve gets drunk
Action is Fred gets drunk
Action is Adam starts to mutually fancy Eve
Action is Fred starts to mutually fancy Eve

Who would you like to partner: Adam or Fred?
Given that you fancy both and they both fancy you.
```
Adam
```
You have chosen Adam, you and Adam are now partners.
As a result of your choice, Fred fancies you less.

Action is Joe begins to disapprove of Eve's relationship with
Adam.

Your friend Joe disapproves of your relationship with Adam.
```

This is because of the age difference: you are teen and Adam is young.
Would you like to leave your partner Adam out of respect for your friendship with Joe?
Yes
You have chosen to leave your partner Adam because your friend Joe disapproved of the relationship.
You and Adam are no longer partners.
Joe now holds your friendship in higher esteem, but Adam no longer fancies you.

Action is Fred starts to fancy Eve
Action is Eve starts to go out with Fred
Action is Eve becomes pregnant with Fred's baby

You are pregnant with Fred's baby. Would you like to keep the baby?
No
You have chosen not to keep the baby. You are no longer pregnant

You are going out with Fred.
Would you like to cheat on Fred with John who you fancy and who fancies you?
No
You have chosen not to cheat on your partner.
As a result of your choice John fancies you less.

Action is Joe encourages you to betray your principle of not stealing.

Your friend Joe is encouraging you to betray your principle of not stealing.
Would you like to betray this principle to join them in stealing?
No
You have chosen not to betray your principle.
As a result of this choice Joe likes you less.

Computational Intelligence in Racing Games

Julian Togelius, Simon M. Lucas and Renzo De Nardi

Department of Computer Science
University of Essex
Colchester CO4 3SQ, UK
{jtogel, sml, rdenar}@essex.ac.uk

Abstract. This chapter surveys the research of us and others into applying evolutionary algorithms and other forms of computational intelligence to various aspects of racing games. We first discuss the various roles of computational intelligence in games, and then go on to describe the evolution of different types of car controllers, modelling of players' driving styles, evolution of racing tracks, comparisons of evolution with other forms of reinforcement learning, and modelling and controlling physical cars. It is suggested that computational intelligence can be used in different but complementary ways in racing games, and that there is unrealised potential for cross-fertilisation between research in evolutionary robotics and CI for games.

1 On the Roles of Computational Intelligence in Computer Games

Computational Intelligence (henceforth "CI") is the study of a diverse collection of algorithms and techniques for learning and optimisation that are somehow inspired by nature. Here we find evolutionary computation, which uses Darwinian survival of the fittest for solving problems, neural networks, where various principles from neurobiology are used for machine learning, and reinforcement learning, which borrows heavily from behaviourist psychology. Such techniques have successfully tackled many complex real-world problems in engineering, finance, the natural sciences and other fields.

Recently, CI researchers have started giving more attention to computer games. Computer games as tools and subjects for research, that is. There are two main (scientific) reasons for this: the idea that CI techniques can add value and functionality to computer games, and the idea that computer games can act as very good testbeds for CI research. Some people (like us) believe that computer games can provide ideal environments in which to evolve complex general intelligence, while other researchers focus on the opportunities competitive games provide for comparing different CI techniques in a meaningful manner.

We would like to begin this paragraph by saying that conversely, many game developers have recently started giving more attention to CI research and incorporated it in their games. However, this would be a lie. At least in commercial games, CI techniques are conspicuously absent, something which CI researchers interested in seeing their favourite algorithms doing some real work often attribute to the game industry's alleged conservatism and risk aversion. Some game developers, on the other hand, claim that the output of the academic Computational Intelligence and Games ("CIG") community is not mature enough (the algorithms are too slow or unreliable), and more importantly, that the research effort is spent on the wrong problems. Game developers and academics appear to see quite different possibilities and challenges in CI for games.

With that in mind, we will provide a brief initial taxonomy of CIG research in this section. Three approaches to CI in games will be described: optimisation, innovation and imitation. The next section will narrow the focus down to racing games, and survey how each of these three approaches have been taken by us and others in the context of such games. In the rest of the chapter, we will describe our own research on CI for racing games in some detail.

1.1 Optimisation

Most CIG research takes the optimisation approach. This means that some aspect of a game is seen as an optimisation problem, and an optimisation algorithm is brought to bear on it. Anything that can be expressed as an array of parameters and where success can in some form be measured can easily be cast as presenting an optimisation problem. The parameters might be values of board positions in a board game, relative amount of resource gathering and weapons construction in a real-time strategy game, personality traits of a non-player character in an adventure game, or weight values of a neural network that controls an agent in just about any kind of game. The optimisation algorithm can be a global optimiser like an evolutionary algorithm, some form of local search, or any kind of problem-specific heuristic. Even CI techniques which are technically speaking not optimisation techniques, such as temporal difference learning (TD-learning), can be used. The main characteristic of the optimisation approach is that we know in advance what we want from the game (e.g. highest score, or being able to beat as many opponents as possible) and that we use the optimisation algorithm to achieve this.

In the literature we can find ample examples of this approach taken to different types of games. Fogel [1] evolved neural networks for board evaluation in chess, and Schraudolph [2] similarly optimised board evaluation functions, but for the game Go and using TD-learning; Runarsson and Lucas [3] compared both methods. Moving on to games that actually require a computer to play (computer games proper, rather than just computerised games) optimisation algorithms have been applied to many simple arcade-style games such as Pacman [4], X-pilot [5] and Cellz [6]. More complicated games for which evolutionary computation has been used to optimise parameters include first-person shooters such as Counter-Strike [7], and real-time strategy games such as Warcraft [8, 9].

Beyond Optimization There is no denying that taking the optimisation approach to CI in games is very worthwhile and fruitful for many purposes. The opportunity many games present for tuning, testing and comparison of various CI techniques is unrivalled. But there are other reasons than that for doing CIG research. Specifically, one might be interested in doing synthetic cognitive science, i.e. studying the artificial emergence of adaptive behaviour and cognition, or one might be interested in coming up with applications of CIG that work well enough to be incorporated in actual commercial computer games. In these cases, we will likely need to look further than optimisation.

From a cognitive science point of view, it is generally less interesting to study the optimal form of a particular behaviour whose general form has been predetermined, than it is to study the emergence (or non-emergence) of new and different behaviours. Furthermore, something so abstract and disembodied as an optimal set of construction priorities in an RTS game doesn't seem to tell us much about cognition; some cognitive scientists would argue that trying to analyse such "intelligence" leads to the *symbol grounding problem*, as the inputs and outputs of the agent is not connected to the (real or simulated) world but only to symbols whose meaning has been decided by the human designer [10].

From the point of view of a commercial game designer, the problem with the optimisation approach is that for many types of games, there isn't any need for better-performing computer-controller agents. In most games, the game can easily enough beat any human player. At least if the game is allowed to cheat a bit, which is simple and invisible enough to do. Rather, there is a need for making the agents' behaviour (and other aspects of the game) more *interesting*.

1.2 Innovation

The border between the optimisation and innovation approaches is not clear-cut. Basically, the difference comes down to that in the optimisation approach we know what sort of behaviour, configuration or structure we want, and use CI techniques to achieve the desired result in an optimal way. Taking the innovation approach, we don't know exactly what we are looking for. We might have a way of scoring or rewarding good results over bad, but we are hoping to create lifelike, complex or just generally interesting behaviours, configurations or structures rather than optimal ones. Typically an evolutionary algorithm is used, but it is here treated more like a tool for search-based design than as an optimiser.

If what is evolved is the controller for an agent, an additional characteristic of the innovation approach is that a closed sensorimotor loop is desired:

where possible, the controller acts on sensor inputs that are "grounded" in the world, and not on pre-processed features or categorisations of the environment, and outputs motor commands rather than abstract parameters to be processed by some hard-coded behaviour mechanism [10]. To make this point about concreteness a bit more concrete, a controller for an first-person shooter that relies on pure (simulated) visual data, and outputs commands to the effect of whether to move left or right and raise or lower the gun, would count as part of a closed sensorimotor loop; a controller that takes as inputs categorizations of situations such as "indoors" and "close to enemies" and outputs commands such as "hide" and "frontal assault" would not. The point is that the perceptions and possible behaviours is influenced and limited by human preconceptions to a much larger extent in the latter example than in the former.

One of the most innovative examples of the innovation approach in a complete computer game is Stanley et al.'s NERO game. Here, real-time evolution of neural networks provide the intelligence for a small army of infantry soldiers [11] The human player trains the soldiers by providing various goals, priorities and obstacles, and the evolutionary algorithm creates neural networks that solve these tasks. Exactly how the soldiers solve the tasks is up to the evolutionary algorithm, and the game is built up around this interplay between human and machine creativity.

1.3 Imitation

In the imitation approach, supervised learning is used to imitate behaviour. What is learned could be either the player's behaviour, or the behaviour of another game agent. While supervised learning is a huge and very active research topic in machine learning with many efficient algorithms developed, this does not seem to have spawned very much research in the imitation approach to CIG. But there are some notable examples among commercial games, such as the critically acclaimed Black and White by Lionhead Studios. In this game, the player takes the role of a god trying to influence the inhabitants of his world with the help of a powerful creature. The creature can be punished and rewarded for its actions, but will also imitate the player's action, so that if the player casts certain spells in certain circumstances, the creature will try to do the same to see whether the player approves of it. Another prominent example of imitation in a commercial game is Forza Motorsport, to be discussed below.

1.4 Are Computer Games the Way Forward for Evolutionary Robotics?

When taken towards mobile robotics rather than games problems, the innovation approach is called evolutionary robotics (ER). This research field, which is 15 years or so old, aims to use evolutionary algorithms to develop as complex and functional robot controllers as possible while incorporating as little human domain knowledge and as few assumptions as possible [12]. The problem with evolutionary robotics is that the most complex and functional robot controllers developed so far are just not very impressive. Common examples of tasks solved successfully are on the order of complexity of determining whether and how to approach one light source or the other based on the feedback gathered last time a light source was approached; similar to the type of problems Skinner subjected his rats to in his (in)famous Skinner boxes. Even very recent examples in collective evolutionary robotics [13, 14], are limited to the evolution of simple consensus strategies or minimal signalling. The evolved controllers, most often neural networks, are on the order of complexity of ten or twenty neurons, which makes for a rather diminutive cognitive faculty compared to those of garden snails and house flies. Not to mention those of yourself or the authors.

There are probably many reasons for this failure of ER to scale up, and we will be content with mention those we think are the most important, without going into detail. First of all, robotics hardware is expensive and slow, severely limiting the applicability of evolutionary algorithms, which often need thousands of fitness evaluations to get anywhere and might well destroy the robot in the course of its exploration of fitness space. Therefore, much evolutionary robotics research go on in simulation. However, the simulators are often not good enough: the evolved robot controllers don't transfer well to the real robots, and the environments, tasks, and sensor inputs are not complex enough. Robots in ER typically have low-dimensional sensors, and it can be argued that complex sensor inputs are necessary for the development of real intelligence (for example, Parker argues that complex ecosystems only came into being with the development of vision during the Cambrian Explosion [15]). And without environments that are complex enough and tasks that exhibit a smooth learning curve, allowing the evolutionary process to "bootstrap" by rewarding the controller for those initial attempts that are less bad than others as well as for truly excellent performance, complex general intelligence won't evolve.

Some already existing computer games might very conceivably provide the right environments for the evolution of complex general intelligence, and the key to scaling up evolutionary robotics. Obviously, controlling agents in computer games is orders of magnitude faster and cheaper than controlling physical robots. Further, hundreds of man-years have gone into the design of many modern commercial games, providing high-resolution graphics (allowing complex sensor inputs), large complex environments, just scoring systems (allowing for good fitness functions) and tasks that are easy to start learning but ultimately require complex general intelligence to solve. Especially relevant in this context is the thinking of game designer Raph Koster, who claims that games are fun to play because we learn from them, and that the most entertaining games are those that teach us best [16]. If he is right, then some computer games are virtually tailor-made for CI research and especially attempts at evolving complex general intelligence, and it would almost be irresponsible not to use them for that.

2 On the Use of Computational Intelligence in Racing Games

Racing games are computer games where the goal is to guide some sort of vehicle towards a goal in an efficient manner. In its simplest form, the challenge for a player comes from controlling the dynamics of the vehicle while planning a good path through the course. In more complicated forms of racing, additional challenge comes from e.g. avoiding collisions with obstacles, shifting gears, adapting to shifting weather conditions, surfaces and visibilities, following traffic rules and last but not least outwitting competing drivers, whether this means overtaking them, avoiding them or forcing them off the track. It is thus possible for a racing game to have a smooth learning curve, in that learning to steer a slow-moving car left or right so as not to crash into a wall is not particularly hard, but on the other side of the complexity continuum, beating Schumacher or Senna at their own game seems to require a lifetime of training. Together with the potential availability of and requirement for high-dimensional sensor data, this means that it should be possible to evolve complex and relatively general intelligence in a racing game.

Let us now look at how CI can be applied to racing games:

2.1 Optimisation

Several groups of researchers have taken this approach towards racing games. Tanev [17] developed an anticipatory control algorithm for an R/C racing simulator, and used evolutionary computation to tune the parameters of this algorithm for optimal lap time. Chaperot and Fyfe [18] evolved neural network controllers for minimal lap time in a 3D motocross game, and we previously ourselves investigated which controller architectures are best suited for such optimisation in a simple racing game [19]. Sometimes optimisation is multiobjective, as in our previous work on optimising controllers for performance on particular racing tracks versus robustness in driving on new tracks [20]. We will discuss this work further in sections 4 and 5.1. And there are other things than controllers that can be optimised in car racing, as is demonstrated by the work of Wloch and Bentley, who optimised the parameters for simulated Formula 1 cars in a physically sophisticated racing game [21], and by Stanley et al., who evolved neural networks for crash prediction in simple car game [22].

2.2 Innovation

In car racing we can see examples of the innovation approach to computational intelligence in work done by Floreano et al. [23] on evolving active vision, work which was undertaken not to produce a controller which would follow optimal race-lines but to see what sort of vision system would emerge from the evolutionary process. We have ourselves studied the effect of different fitness measures in competitive co-evolution of two cars on the same tracks, finding that qualitatively different behaviour can emerge depending on whether controllers are rewarded for relative or absolute progress [24], work which will be described in section 5.2. Proving that the innovation approach can be applied to other aspects of racing games than controllers, we have worked on evolving racing tracks for enhanced driving experience for particular human players [25, 26], work which will be presented in section 8.

2.3 Imitation

A major example of the imitation approach to computational intelligence in racing games is the XBox game Forza Motorsport from Microsoft Game Studios. In this game, the player can train a "drivatar" to play just like himself, and then use this virtual copy of himself to get ranked on tracks he doesn't want to drive himself, or test his skill against other players' drivatars. Moving from racing games to real car driving, Pomerleau's work on teaching a real car to drive on highways through supervised learning based on human driving data is worth mentioning as an example of the imitation approach [27]. The reason for using imitation rather than optimisation in this case was probably not that interesting driving was preferred to optimal driving, but rather that evolution using real cars on real roads would be costly. Our own work on imitating the driving styles of human players was published in [25] and [26] and is discussed in section 7.

3 Our Car Racing Model

The experiments described below were performed in a 2-dimensional simulator, intended to qualitatively if not quantitatively, model a standard radio-controlled (RC) toy car (approximately 17 cm long) in an arena with dimensions approximately 3 × 2 meters, where the track is delimited by solid walls. The simulation has the dimensions 400 × 300 $pixels$, and the car measures 20 × 10 $pixels$.

RC toy car racing differs from racing full-sized cars in several ways. One is the simplified, "bang-bang" controls: many toy RC cars have only three possible drive modes (forward, backward, and neutral) and three possible steering modes (left, right and center). Other differences are that many toy cars have bad grip on many surfaces, leading to easy skidding, and that damaging such cars in collisions is nigh impossible due to their low weight.

In our model, a track consists of a set of walls, a chain of waypoints, and a set of starting positions and directions. When a car is added to a track in one of the starting positions, with corresponding starting direction, both the position and angle being subject to random alterations. The waypoints are used for fitness calculations.

The dynamics of the car are based on a reasonably detailed mechanical model, taking into account the small size of the car and bad grip on the surface, but is not based on any actual measurement [28, 29]. The collision response system was implemented and tuned so as to make collisions reasonably realistic after observations of collisions with our physical RC cars. As an effect, collisions are generally undesirable, as they may cause the car to get stuck if the wall is struck at an unfortunate angle and speed.

Still, there are differences between our model and real RC cars. Obviously, parameters such as acceleration and turning radius don't correspond exactly to any specific RC car and surface combination, as is the case with the skidding behaviour. A less obvious difference is that our model is symmetric in the left/axis, while most toy cars have some sort of bias here. Another difference has to do with sensing: as reported in Tanev et al. as well as [19], if an overhead camera is used for tracking a real car, there is a small but not unimportant lag in the communication between camera, computer and car, leading to the controller acting on outdated perceptions. Apart from that, there is often some error in estimations of the car's position and velocity from an overhead camera. (Though this is significantly less of a problem if using a more sophisticated motion capture system, as discussed in section 9.) In contrast, the simulation allows instant and accurate information to be fed to the controller.

4 Evolving Controllers for the Basic Racing Task

In our first paper on evolutionary car racing, we investigated how best to evolve controllers for a single car racing around a single track as fast as possible [19]. Our focus was on what sort of information about the car's state and environment was needed for effective control, and how this information and the controller should be represented. The fitness function and evolutionary algorithm was kept fixed for all experiments, and is described below. We also discuss which approaches turned out to work and which didn't, and what we can learn from this.

4.1 Evolutionary Algorithm, Neural Networks, and Fitness Function

The evolutionary algorithm we used in these experiments, which is also the basis for the algorithm used in all subsequent experiments in this chapter, is a 50+50 evolution strategy. It's workings are fairly simple: at the beginning of an evolutionary run, 100 controllers are initialised randomly. Each generation, each controller is tested on the task at hand and assigned a fitness value, and the population is sorted according to fitness. The 50 worst controllers are then discarded, and replaced with copies of the 50 best controllers. All the new controllers are mutated, meaning that small random changes are made to them; exactly what these random changes consist in is dependent on the controller architecture. For the experiments in this chapter, most evolutionary runs lasted 100 generations.

Three of the five architectures in this section, including the winning architecture, are based on neural networks. The neural networks are standard multi-layer perceptrons, with n input neurons, a single layer of h hidden neurons, and two output neurons, where each neuron implements the *tanh* transfer function. At each time step, the inputs as specified by the experimental setup is fed to the network, activations are propagated, and the outputs of the network are interpreted as actions that are used to control the car. Specifically, an activation of less than -0.3 of output 0 is interpreted as backward, more than 0.3 as forward and anything in between as no motor action; in the same fashion, activations of output 1 is interpreted as steering left, center or right.

At the start of an evolutionary run, the $m \times n \times 2$ connections are initialized to strength 0. The mutation operator then works by applying a different random number, drawn from a gaussian distribution around zero with standard deviation 0.1, to the strength of each connection.

As for the fitness calculation, a track was constructed to be qualitatively similar to a tabletop racing track we have at the University of Essex (The real and model tracks can be seen in figures 1 and 2). Eleven waypoints were defined, and fitness was awarded to a controller proportionally to how many waypoints it managed to pass in one trial. The waypoints had to be passed in order, and the car was considered to have reached a waypoint when its centre was within 30 *pixels* of the centre of the waypoint. Each trial started with the car in a pre-specified starting position and orientation, lasted for 500 time steps, and awarded a fitness of 1 for a controller that managed to drive exactly one lap of the track within this time (0 for not getting anywhere at all, 2 for two laps and so on in increments of 1/11). Small random perturbations were applied to starting positions and all sensor readings.

Fig. 1. The tabletop RC racing setup

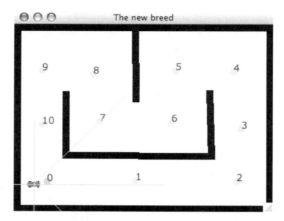

Fig. 2. The simulated racing track with the car in the starting position. The light gray lines represent wall sensors

4.2 How not to Evolve Good Driving

Four controller architectures were tried and found not to work very well, or not to work at all. These were action sequences, open-loop neural networks, neural networks with third-person information, and force fields.

Action Sequences Our action sequences were implemented as one-dimensional arrays of length 500, containing actions, represented as integers in the range 0–8. An action is a combination of driving command (forward, backward, or neutral) and steering commands (left, right or center). When evaluating an action sequence controller, the car simulation at each time step executes the action specified at the corresponding index in the action sequence. At the beginning of each evolutionary run, controllers are initialized as sequences of zeroes. The mutation operator then works by selecting a random number of positions between 0 and 100, and changing the value of so many positions in the action sequence to a new randomly selected action.

The action sequences showed some evolvability, but we never managed to evolve a controller which could complete a full lap; best observed fitness was around 0.5. Often, the most successful solutions consisted in the car doing virtually nothing for the better part of the 500 time steps, and then suddenly shooting into action towards the end. This is because each action is associated with a time step rather than a position on the track, so that a mutation early in the sequence that is in itself beneficial (e.g. accelerating the car at the start of a straight track section) will offset the actions later in the sequence in such a way that it probably lowers the fitness as a whole, and is thus selected against. As an action sequence can represent any behaviour, this controller is obviously not limited in its representational power, but apparently in its evolvability.

Computational Intelligence in Racing Games   49

Open-loop Neural Networks Another attempt to evolve open-loop controllers was made with neural networks. These neural networks took only the current time step divided by the total number of time steps as input, and outputs were interpreted as above. The intuition behind this was that as multi-layer perceptrons have been theoretically shown to approximate any function with arbitrary precision, this would be the case with a function from time step to optimal action. However, we didn't manage to evolve any interesting behaviour at all with this setup, indeed very few action changes were observed, and very few waypoints passed. Another case of practical nonevolvability.

Neural Networks with Third-person Information Our next move was to add six inputs to the neural network describing the state of the car at the current timestep: the x and y components of the car's position, the x and y components of its velocity, its speed and its orientation. All inputs were scaled to be in the range -10 to 10. Despite now having an almost complete state description, these controllers evolved no better than the action sequences.

Force Fields Force field controllers are common in mobile robotics research ([30]), and we wanted to see whether this approach could be brought to bear on racing cars, with all their momentum and non-holonomicity. A force field controller is here defined as a two-dimensional array of two-tuples, describing the preferred speed and preferred orientation of the car while it is in the field. Each field covers an area of $n \times n$ *pixels*, and as the fields completely tile the track without overlapping, the number of fields are $\frac{l}{n}\frac{w}{n}$, where l is length, and w is width of the track, respectively. At each time-step, the controller finds out which field the centre of the car is inside, and compares the preferred speed and orientation of that field with the cars actual speed and orientation. If the actual speed is less than the preferred, the controller issues an action containing a forward command, otherwise it issues a backward command; if the actual orientation of the car is left of the preferred orientation, the issued action contains a steer right command, otherwise it steers left. We tried various parameters but the ones that seemed least bad were fields with the size 20×20 pixels, evolved with gaussian mutation of all fields with magnitude 0.1. However, these controllers fared no better than the neural networks with third-person information. Some of the best evolved force-field controllers go half the way around the track, before the car gets stuck between two force fields, going forwards and backwards forever.

4.3 How to Evolve Good Driving

The approach that actually worked well was based on neural networks and simulated range-finder wall sensors. In this experimental setup, the five inputs to the neural network consisted of the speed of the car and the outputs of three wall sensors and one aim point sensor. The aim point sensor simply outputs the difference between the car's orientation and the angle from the center of the car to the next aim point, yielding a negative value if that point is to the left of the car's orientation and a positive value otherwise.

Each of the three wall sensors is allowed any forward facing angle (i.e. a range of 180 *degrees*), and a reach, between 0 and 100 *pixels*. These parameters are co-evolved with the neural network of the controller. The sensor works by checking whether a straight line extending from the centre in the car in the angle specified by that sensor intersects with the wall at eleven points positioned evenly along the reach of the sensor, and returning a value equal to 1 divided by the position along the line which first intersects a wall. Thus, a sensor with shorter reach has higher resolution, and evolution has an incentive to optimize both reaches and angles of sensors. Gaussian random noise with standard deviation 0.01 is also added to each of the sensor readings.

Results of these experiments were excellent. Within 100 generations, good controllers with fitness of at least 2 (i.e., that managed to drive at least two laps within the allotted time) were reliably evolved, and the best controllers managed to reach fitness 3.2. We also compared the networks' performance to humans controlling the car with the keyboard, and the best evolved controller out of 10 runs narrowly beat the best human driver out of 10 human players who tried their luck at the game at the 2005 IEEE Symposium on Computational Intelligence and Games demonstration session.

A couple of variations of this setup was tested in order to explore its robustness. It was found that controllers could be evolved that reliably took the car around the track without using the waypoint sensor, solely relying on wall sensors and speed. However, these controllers drove slower, achieving fitness values around 2. (If, on the other hand, the waypoint sensor input was removed from controllers that were evolved to use it, their fitness quickly dropped to zero.) Evolving proficient controllers that used only waypoint sensor and no wall sensors was only possible when all randomness was removed from the simulation.

4.4 Take-home Lessons

Our first experiments clearly showed that the representation of the car's state and environment is crucial to whether a good controller can be evolved. Several controller architectures that should in theory be adequate for controlling the car around a single track proved to be nonevolvable in practice due to the lack of first-person sensor data. Interestingly, the controller architecture that won out is also the one which is arguably the most interesting to study from a cognitive science perspective, acting on relatively "raw" sensor data that would be trivially available to a suitably equipped real RC car. Subsequent unpublished experiments have indicated that the representation of the controller itself is not nearly as important as the format of the sensor data; we have managed to achieve proficient control using both genetic programming, and through evolving tables of actions associated with points (cluster centres) in sensor-data space. In both these cases we used the same sensor setup as above.

Another important thing to note is that the existence of impermeable walls significantly changes the problem. Letting the car drive through walls results in a much simpler problem, as could be expected.

5 Evolving Controllers for More Complicated Tasks

The above experiments only concern one car at a time, starting from the same position on the same track. However, car racing gets really interesting (and starts demanding a broader range of cognitive skills) when several cars compete on several racing tracks. The good news is that the nature of the racing task is such that it is possible to gradually increase the complexity level from single-car single-track race to multi-car multi-track races.

For the experiments in this section we designed eight different tracks, presented in figure 3. The tracks are designed to vary in difficulty, from easy to hard. Three of the tracks are versions of three other tracks with all the waypoints in reverse order, and the directions of the starting positions reversed. Various minor changes were also made to the simulator between the above and below experiments (e.g. changing the collision model, and changing the number of time steps per trial to 700), which means that fitnesses cannot be directly compared between these experiments.

5.1 Generalisation and Specialisation

First, we focused on single-car, multi-track racing. In these experiments, each controller was equipped with six wall-sensors, and also received speed and waypoint sensor inputs. In some of the experiments, the position and ranges of the sensors were evolvable, in some they were fixed, as illustrated in fig. 4.

Evolving Track-specific Controllers The first experiments consisted in evolving controllers for the eight tracks separately, in order to rank the difficulty of the tracks. For each of the tracks, the evolutionary algorithm was run 10 times, each time starting from a population of "clean" controllers, with all connection weights set to zero and sensor parameters as explained above. Only weight mutation was allowed. The evolutionary runs were for 200 generations each.

The results are listed in table 1, which is read as follows: each row represents the results for one particular track. The first column gives the mean of the fitnesses of the best controller of each of the evolutionary runs at generation 10, and the standard deviation of the fitnesses of the same controllers. The next three columns present the results of the same calculations at

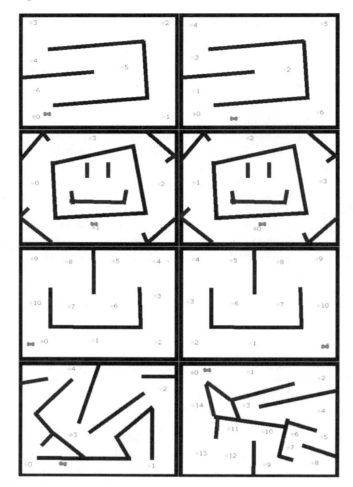

Fig. 3. The eight tracks. Notice how tracks 1 and 2 (at the top), 3 and 4, 5 and 6 differ in the clockwise/anti-clockwise layout of waypoints and associated starting points. Tracks 7 and 8 have no relation to each other apart from both being difficult

Fig. 4. The fixed sensor configuration for the incremental evolution experiments

Table 1. The fitness of the best controller of various generations on the different tracks, and number of runs producing proficient controllers. Fitness averaged over 10 separate evolutionary runs; standard deviation between parentheses

Track	10	50	100	200	Pr.
1	0.32 (0.07)	0.54 (0.2)	0.7 (0.38)	0.81 (0.5)	2
2	0.38 (0.24)	0.49 (0.38)	0.56 (0.36)	0.71 (0.5)	2
3	0.32 (0.09)	0.97 (0.5)	1.47 (0.63)	1.98 (0.66)	7
4	0.53 (0.17)	1.3 (0.48)	1.5 (0.54)	2.33 (0.59)	9
5	0.45 (0.08)	0.95 (0.6)	0.95 (0.58)	1.65 (0.45)	8
6	0.4 (0.08)	0.68 (0.27)	1.02 (0.74)	1.29 (0.76)	5
7	0.3 (0.07)	0.35 (0.05)	0.39 (0.09)	0.46 (0.13)	0
8	0.16 (0.02)	0.19 (0.03)	0.2 (0.01)	0.2 (0.01)	0

generations 50, 100 and 200, respectively. The "Pr" column gives the number of proficient best controllers for each track. An evolutionary run is deemed to have produced a proficient controller if its best controller at generation 200 has a fitness (averaged, as always, over three trials) of at least 1.5, meaning that it completes at least one and a half lap within the allowed time.

For the first two tracks, proficient controllers were produced by the evolutionary process within 200 generations, but only in two out of ten runs. This means that while it is possible to evolve neural networks that can be relied on to race around one of these track without getting stuck or taking excessively long time, the evolutionary process in itself is not reliable. In fact, most of the evolutionary runs are false starts. For tracks 3, 4, 5 and 6, the situation is different as at least half of all evolutionary runs produce proficient controllers. The best evolved controllers for these tracks get around the track fairly fast without colliding with walls. For tracks 7 and 8, however, we have not been able to evolve proficient controllers from scratch at all. The "best" (least bad) controllers evolved for track 7 might get halfway around the track before getting stuck on a wall, or losing orientation and starting to move back along the track.

Evolving controllers from scratch with sensor parameter mutations turned on resulted in somewhat lower average fitnesses and numbers of proficient controllers.

Generality of Evolved Controllers Next, we examined the generality of these controllers by testing their performance of the best controller for each track on each of the ten tracks. The results are presented in Table 2, and clearly show that the generality is very low. No controller performed very well on any track it had not been evolved on, with the interesting exception of the controller evolved for track 1, that actually performed better on track 3 than on the track for which it had been evolved, and on which it had a rather mediocre performance. It should be noted that both track 1 and track 3 (like all odd-numbered tracks) run counter-clockwise, and there indeed seems to be

Table 2. The fitness of each controller on each track. Each row represents the performance of the best controller of one evolutionary run with fixed sensors, evolved the track with the same number as the row. Each column represents the performance of the controllers on the track with the same number as the column. Each cell contains the mean fitness of 50 trials of the controller given by the row on the track given by the column. Cells with bold text indicate the track on which a certain controller performed best

$Evo/Test$	1	2	3	4	5	6	7	8
1	1.02	0.87	**1.45**	0.52	1.26	0.03	0.2	0.13
2	0.28	**1.13**	0.18	0.75	0.5	0.66	0.18	0.14
3	0.58	0.6	**2.1**	1.45	0.62	0.04	0.03	0.14
4	0.15	0.32	0.06	**1.77**	0.22	0.13	0.07	0.13
5	0.07	-0.02	0.05	0.2	**2.37**	0.1	0.03	0.13
6	1.33	0.43	0.4	0.67	1.39	**2.34**	0.13	0.14
7	**0.45**	0	0.6	0.03	0.36	0.07	0.22	0.08
8	0.16	0.28	0.09	**0.29**	0.21	0.08	0.1	0.13

a slight bias for the other controllers to get higher fitness on tracks running in the same direction as the track for which they were evolved.

Controllers evolved with evolvable sensor parameters turn out to generalize about as badly as the controllers evolved with fixed sensors.

Evolving Robust General Controllers The next suite of experiments were on evolving robust controllers, i.e. controllers that can drive proficiently on a large set of tracks. First, we attempted to do this by simultaneous evolution: starting from scratch (networks with all connection weights set to zero), each controller was tested on all the first six tracks, and its fitness on these six tracks was averaged. This proved not to work at all: no controllers evolved to do anything more interesting than driving straight forward. So we turned to incremental evolution.

The idea here was to evolve a controller on one track, and when it reached proficiency (mean fitness above 1.5) add another track to the training set - so that controllers are now evaluated on both tracks and fitness averaged - and continue evolving. This procedure is then repeated, with a new track added to the fitness function each time the best controller of the population has an average fitness of 1.5 or over, until we have a controller that races all of the first six tracks proficiently. The order of the tracks was 5, 6, 3, 4, 1 and finally 2, the rationale being that the balance between clockwise and counterclockwise should be as equal as possible in order to prevent lopsided controllers, and that easier tracks should be added to the mix before harder ones.

This approach turned out to work much better than simultaneous evolution. Several runs were performed, and while some of them failed to produce generally proficient controllers, some others fared better. A successful run usually takes a long time, on the order of several hundred generations, but

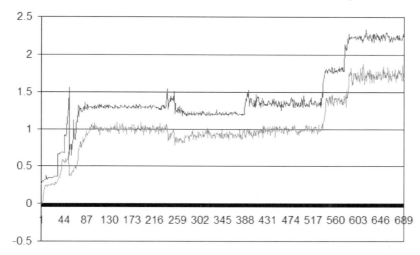

Fig. 5. A successful run of the incremental evolution of general driving skill

it seems that once a run has come up with a controller that is proficient on the first three or four tracks, it almost always proceeds to produce a generally proficient controller. One of the successful runs is depicted in figure 5 and the mean fitness of the best controller of that run when tested on all eight tracks separately is shown in 3. As can be seen from this table, the controller does a good job on the six tracks for which it was evolved, bar that it occasionally gets stuck on a wall in track 2. It never makes its way around track 7 or 8.

These runs were all made with sensor mutation turned off. Turning on sensor mutation at the start of a run of incremental evolution seems to lead to "premature specialisation", where the sensor configuration becomes specialised to some tracks and can't be used to control the car on other tracks. Evolving sensor parameters can be beneficial, however, when this is done for a controller that has already reached general proficiency. In our tests this further evolution added between 0.1 and 0.2 to the average fitness of the controller over the first six tracks.

Specialising Controllers In order to see whether we could create even better controllers, we used one of the further evolved controllers (with evolved sensor parameters) as basis for specializing controllers. For each track, 10 evolutionary runs were made, where the initial population was seeded with the general controller and evolution was allowed to continue for 200 generations. Results are shown in table 3. The mean fitness improved significantly on all six first tracks, and much of the fitness increase occurred early in the evolutionary run. Further, the variability in mean fitness of the specialized controllers from different evolutionary runs is very low, meaning that the reliability of the evolutionary process is very high. Perhaps most surprising, however, is

Table 3. Fitness of best controllers, evolving controllers specialised for each track, starting from a further evolved general controller with evolved sensor parameters

$Track$	10	50	100	200	$Pr.$
1	1.9 (0.1)	1.99 (0.06)	2.02 (0.01)	2.04 (0.02)	10
2	2.06 (0.1)	2.12 (0.04)	2.14 (0)	2.15 (0.01)	10
3	3.25 (0.08)	3.4 (0.1)	3.45 (0.12)	3.57 (0.1)	10
4	3.35 (0.11)	3.58 (0.11)	3.61 (0.1)	3.67 (0.1)	10
5	2.66 (0.13)	2.84 (0.02)	2.88 (0.06)	2.88 (0.06)	10
6	2.64 (0)	2.71 (0.08)	2.72 (0.08)	2.82 (0.1)	10
7	1.53 (0.29)	1.84 (0.13)	1.88 (0.12)	1.9 (0.09)	10
8	0.59 (0.15)	0.73 (0.22)	0.85 (0.21)	0.93 (0.25)	0

that all 10 evolutionary runs produced proficient controllers for track 7, on which the general controller had not been trained (and indeed had very low fitness) and for which it had previously been found to be impossible to evolve a proficient controller from scratch.

Take-home Lessons First of all, these experiments show that it is possible to evolve controllers that display robust and general driving behaviour. This is harder to achieve than good driving on a single track, but can be done using incremental evolution, where the task to be performed gradually grows more complex. This can be seen as some very first steps towards incrementally evolving complex general intelligence in computer games. Our other main result in this section, that general controllers quickly and reliably can be specialised for very good performance on particular tracks, is good news for the applicability of these techniques in computer games. For example, this method could be used to generate drivers with particular characteristics on the fly for a user-generated track.

5.2 Competitive Co-evolution

So, how about multi-car races? Our next set of experiments was concerned with how to generate controllers that drive well, or at least interestingly, in the presence of another car on the same track [24]. For this purpose, we used tracks 1, 3, and 5 from figure 3, which all run clockwise and for which it is therefore possible to evolve a proficient controller from scratch using simultaneous evolution. Each controller was evaluated on all three tracks.

Solo-evolved Controllers Together In order to characterise the difficulty of the problem, the natural first step was to place two cars, controlled by different general controllers developed using the incremental approach above, on the same track at the same time, with starting positions relatively close to each other. These two cars have only wall sensors, waypoint sensor and speed inputs, and so have no knowledge of each other's presence. What happens when they are unleashed on the same track is that they collide with each other, and as a direct consequence of this most often either collide with a wall or lose track of which way they were going and start driving backwards. This leads to very low fitnesses. So apparently we need to evolve controllers specifically for competitive driving.

Sensors, Fitness and Evolutionary Algorithm The first step towards co-evolving competitive controllers was equipping the cars with some new sensors. In our co-evolutionary experiments each car has eight range-finder sensors, and the evolutionary process decides how many of them are wall sensors and how many are car sensors. The car sensors work just like the wall sensors, only that they report the approximate distance to the other car if in its line of detection, rather than the closest wall.

Measuring fitness in a co-evolutionary setting is not quite as straightforward as in the single-car experiments above. At least two basic types of fitness can be define: absolute and relative fitness. Absolute fitness is simply how far the car has driven along the track within the allotted time, and is the fitness measure we used in earlier experiments. Relative fitness is how far ahead of the other car the controller's car is at the end of the 700 time steps. As we shall see below, these two fitness measures can be combined in several ways.

The (single-population) co-evolutionary algorithm we used was a modified version of the evolution strategy used for earlier experiments. The crucial difference was that the fitness of each controller was determined by letting it race against three different controllers from the fitter half of the population.

Co-evolution To test the effect of the different fitness measures, 50 evolutionary runs were made, each consisting of 200 generations. They were divided into five groups, depending on the absolute/relative fitness mix used by the selection operator of the co-evolutionary algorithm: ten evolutionary runs were performed with absolute fitness proportions 0.0, 0.25, 0.5, 0.75 and 1.0 respectively. These were then tested in the following manner: the best individuals from the last generation of each run were first tested for 50 trials on all three tracks without competitors, and the results averaged for each group. Then, all five controllers in each group were tested for 50 trials each in competition against each controller of the group. See table 4 for results.

What we can see from this table is that the controllers evolved mainly for absolute fitness on average get further along the track than controllers evolved mainly for relative fitness. This applies both in competition with other cars and when alone on the track, though all controllers get further when they are alone on the track than when they are in competition with another car. Further, the co-evolved cars drive slower and get lower fitness than the solo-evolved controllers. So there are several trade-offs in effect.

What we can't see from the table, but is apparent when looking at the game running, is that the evolved controllers behave wildly differently. Most

Table 4. The results of co-evolving controllers with various proportions of absolute fitness. All numbers are the mean of testing the best controller of ten evolutionary runs for 50 trials. Standard deviations in parentheses

$Proportion absolute$	0.0	0.25	0.5	0.75	1.0
Absolute fitness solo	1.99 (0.31)	2.09 (0.33)	2.11 (0.35)	2.32 (0.23)	2.23 (0.23)
Absolute fitness duo	0.99 (0.53)	0.95 (0.44)	1.56 (0.45)	1.44 (0.44)	1.59 (0.45)
Relative fitness duo	0 (0.75)	0 (0.57)	0 (0.53)	0 (0.55)	0 (0.47)

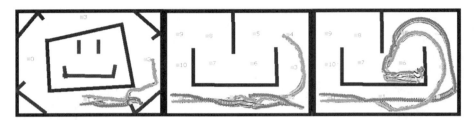

Fig. 6. Traces of the first 100 or so time-steps of three runs that included early collisions. From left to right: red car pushes blue car to collide with a wall; red car fools blue car to turn around and drive the track backwards; red and blue car collide several times along the course of half a lap, until they force each other into a corner and both get stuck. Note that some trials see both cars completing 700 time-steps driving in the right direction without getting stuck

importantly, the controllers evolved mainly for relative fitness are much more aggressive than those evolved mainly for absolute fitness. They typically aim for collisions with the other car whenever the other car stands to lose more from it (e.g. it can be forced to collide with a wall), while those evolved for absolute fitness generally avoid collision at all times. We have put some videos of such interesting car to car interactions on the web, as a video in this case says more than a thousand statistics. ([31]). See also figure 6.

We ran several auxiliary experiments using these evolved controllers, comparing different categories of controllers against each other. The most interesting result out of these comparisons is that solo-evolved controllers on average perform better than co-evolved controllers when one of each kind is placed on the same track. In fact, solo-evolved controllers perform much better against co-evolved controllers than against other solo-evolved controllers for the simple reason that the solo-evolved controllers drive faster and thus don't collide with the other car. We seem to have another trade-off here, between being able to outsmart another car and driving fast. Of course, a truly adaptive controller would recognise at what speed the other car drove and speed up if it was lagging behind, but this is not happening. At the moment we don't know whether the reason this is not happening is the simplicity of our co-evolutionary algorithm or the simplicity of the sensors and neural network we use, but we guess it's a bit of both.

5.3 Take-home Messages

The above example shows that the innovation approach can be taken towards the car-racing problem with good results, in that the co-evolutionary algorithm produced controllers displaying interesting and unexpected behaviour. The inability of our algorithms to produce controllers that were as general and flexible as we would have liked also demonstrates how quickly a seemingly simple problem can be turned into a highly complex problem, which might require anticipatory capabilities and complex sensing to be solved satisfactorily. This is fertile ground for future research.

6 An Application: Personalised Content Creation

In the following two sections, we will look at a possible application in games, building on the results of the above experiments. Our goal here is to automatically generate racing tracks that are fun for a particular player. The method is to model the driving style of the human player, and then evolving a new track according to a fitness function designed to measure entertainment value of the track. How to measure entertainment value is a truly complex topic which we have discussed in more depth in our publications on these experiments [25, 26]. Here we will focus on the technical issues of player modelling and track evolution.

6.1 The Cascading Elitism Algorithm

We use artificial evolution both for modelling players and constructing new tracks, and in both cases we have to deal with multiobjective fitness functions. A simple way to deal with this is to modify our evolution strategy to use multiple elites. In the case of three fitness measures, it works as follows: out of a population of 100, the best 50 genomes are selected according to fitness measure $f1$. From these 50, the 30 best according to fitness measure $f2$ are selected, and finally the best 20 according to fitness measure $f3$ are selected. Then these 20 individuals are copied four times each to replenish the 80 genomes that were selected against, and finally the newly copied genomes are mutated.

This algorithm, which we call Cascading Elitism, is inspired by an experiment by Jirenhed et al. [32].

7 Modelling Player Behaviour

The first step towards generating personalised game content is modelling the driving style of the player. In the context of driving games, this means learning a function from current (and possibly past) state of the car (and possibly environment and other cars) to action, where the action would ideally be the same action as the human player would take in that situation. In other words, we are teaching a controller to imitate a human. As we shall see this can be done in several quite different ways.

7.1 What should be Modelled?

The first question to consider when modelling human driving behaviour is what, exactly, should be modelled. This is because the human brain is much more complex than anything we could currently learn, even if we had the training data available. It is likely that a controller that accurately reproduces the player's behaviour in some respects and circumstances work less well in others. Therefore we need to decide what features we want from the player model, and which features have higher priority than others.

As we want to use our model for evaluating fitness of tracks in an evolutionary algorithm, and evolutionary algorithms are known to exploit weaknesses in fitness function design, the highest priority for our model is robustness. This means that the controller does not act in ways that are grossly inconsistent with the modelled human, especially that it does not crash into walls when faced with a novel situation. The second criterion is that the model has the same average speed as the human on similar stretches of track, e.g. if the human drives fast on straight segments but slows down well before sharp turns, the controller should do the same. That the controller has a similar driving style to the human, e.g. drives in the middle of straight segments but close to the inner wall in smooth curves (if the human does so), is also important but has a lower priority.

7.2 Direct Modelling

The most straightforward way of acquiring a player model is direct modelling: log the state of the car and the action the player takes over some stretch of time, and use some form of supervised learning to associate state with action. Accordingly, this was the first approach to player modelling we tried. Both backpropagation of neural networks and nearest neighbour classification was used to map sensor input (using ten wall sensors, speed and waypoint sensor) to action, but none of these techniques produced controllers that could drive more than half a lap reliably. While the trained neural networks failed to produce much meaningful behaviour at all, the controllers based on nearest neighbour classification sometimes reproduced player behaviour very faithfully, but when faced with situations which were not present in the training data they typically failed too. Typically, failure took the form of colliding with a wall and getting stuck, as the human player had not collided with any walls during training and the sequence of movements necessary to back off from a collision was not in the training data. This points to a more general problem with direct modelling: as long as a direct model is not perfect, its performance will always be inferior (and not just different) to the modelled human. Specifically, direct models lack robustness.

7.3 Indirect Modelling

Indirect modelling means measuring certain properties of the player's behaviour and somehow inferring a controller that displays the same properties. This approach has been taken by e.g. Yannakakis in a simplified version of the Pacman game [33]. In our case, we start from a neural network-based controller that has previously been evolved for robust but not optimal performance over a wide variety of tracks, as described in section 5.1. We then continue evolving this controller with the fitness function being how well its behaviour agrees with certain aspects of the human player's behaviour. This way we satisfy the top-priority robustness criterion, but we still need to decide on what fitness function to employ in order for the controller to satisfy the two other criteria described above, situational performance and driving style.

First of all, we design a test track, featuring a number of different types of racing challenges. The track, as pictured in figure 7, has two long straight sections where the player can drive really fast (or choose not to), a long smooth curve, and a sequence of nasty sharp turns. Along the track are 30 waypoints, and when a human player drives the track, the way he passes each waypoint is recorded. What is recorded is the speed of the car when the waypoint is passed, and the orthogonal deviation from the path between the waypoints, i.e. how far to the left or right of the waypoint the car passed. This matrix of two times 30 values constitutes the raw data for the player model.

The actual player model is constructed using the Cascading Elitism algorithm, starting from a general controller and evolving it on the test track. Three fitness functions are used, based on minimising the following differences between the real player and the controller: f_1: total progress (number of

Fig. 7. The test track used for player modelling

waypoints passed within 1500 timesteps), *f2:* speed at which each waypoint was passed, and *f3:* orthogonal deviation at each waypoint as it was passed. The first and most important fitness measure is thus total progress difference, followed by speed and deviation difference respectively. Using this technique, we successfully modelled two of the authors, verifying by inspection that the acquired controllers drove similarly to the authors in question.

8 Evolving Racing Tracks

Developing a reliable quantitative measure of fun is not exactly straightforward. We have previously reasoned at some length about this, however in the experiments described here we chose a set of features which would be believed not to be too hard to measure, and designed a fitness function based on these. The features we want our track to have for the modelled player, in order of decreasing priority, is the right amount of challenge, varying amount of challenge, and the presence of sections of the track in which it is possible to drive really fast. The corresponding fitness functions are the negative difference between actual progress and target progress (in this case defined as 30 waypoints in 700 timesteps), variance in total progress over five trials of the same controller on the same track, and maximum speed.

8.1 Track Representation

The representation we present here is based on b-splines, or sequences of Bezier curves joined together. Each segment is defined by two control points, and two adjacent segment always share one control point. The remaining two control points necessary to define a Bezier curve are computed in order to ensure that the curves have the same first and second derivatives at the point they join, thereby ensuring smoothness. A track is defined by a b-spline containing 30 segments, and mutation is done by perturbing the positions of their control points.

The collision detection in the car game works by sampling pixels on a canvas, and this mechanism is taken advantage of when the b-spline is transformed into a track. First thick walls are drawn at some distance on each side of the b-spline, this distance being either set to 30*pixels* or subject to evolution depending on how the experiment is set up. But when a turn is too sharp for the current width of the track, this will result in walls intruding on the track and sometimes blocking the way. The next step in the construction of the track is therefore "steamrolling" it, or traversing the b-spline and painting a thick stroke of white in the middle of the track. Finally, waypoints are added at approximately regular distances along the length of the b-spline. The resulting track can look very smooth, as evidenced by the test track which was constructed simply by manually setting the control points of a spline.

8.2 Evolutionary Method

Each evolutionary run starts from an equally spaced radial disposition of the control points around the center of the image; the distance of each point from the center is generated randomly. Mutation works by going through all control points, and adding or subtracting an amount of distance from the center of the track drawn from a gaussian distribution. Constraining the control points in a radial disposition is a simple method to exclude the possibility of producing a b-spline containing loops, therefore producing tracks that are always fully drivable.

In figure 8 two tracks are displayed that are evolved to be fun for two of the authors. One of the authors is a significantly better driver of this particular racing game than the other, having spent too much time in front of it, and so his track has more sharp turns and narrow passages. The track that was optimised for entertaining the other player is smoother and easier to drive. In figure 9 we have plotted the progress, maximum speed, and progress variance over the evolutionary runs that produced the tracks in figure 8.

Finally, figure 10 presents a track that was evolved for the more proficient driver of the two authors using the same representation and mutation, but with a "random drift" initialisation. Here, we first do 700 mutations without selecting for higher fitness, keeping all mutations after which it is still possible for controller to drive at least one lap on the track, and retracting other mutations. After that, evolution proceeds as above.

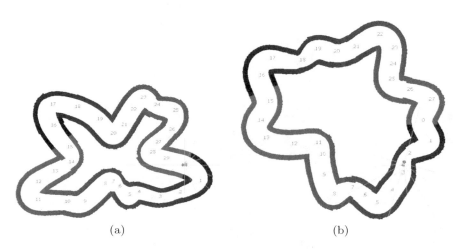

Fig. 8. Track evolved for a proficient player (a), and for a not proficient one (b) using the radial method. Although this method is biased towards "flower-like tracks", is clear that the first track is more difficult to drive, given its very narrow section an its sharp turns. A less proficient controlled instead produced an easy track with gentle turns and no narrow sections

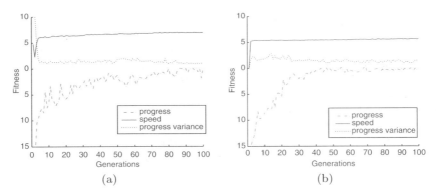

Fig. 9. Fitness graph of the evolution runs that produced the tracks in picture 8. The cascading elitism algorithm clearly optimizes for the final progress and the speed fitness. However given the interplay between progress variation and maximum speed, progress variations ends up being initially sightly reduced. This is a direct result of having speed as second (and therefore more important than the third) fitness measure in the selection process of the cascade

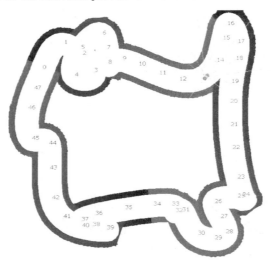

Fig. 10. Track evolved for a proficient driver using the random walk method. This track is arguably less "flower-like" than the tracks evolved without random walk initialisation

9 The Roads Ahead: Models, Reinforcements, and Physical Driving

While the experiments described above show that good driving behaviour can be evolved, with human-competitive performance even on complex tracks, and showcased player modelling and track evolution as a potential application in games, there are several outstanding research questions. The main question is of course exactly how complex driving behaviour it is possible to evolve. Other questions include whether we can evolve controllers for physical RC cars as well as simulated ones, and how well our evolutionary methods compare to other methods of automatically developing controllers.

Let us start with the physical car racing. We have recently set up a system where one of our inexpensive toy RC cars is controlled by a desktop PC through a standard 27 Mhz transmitter modified to work with the PC's parallel port. The car is tracked through infrared reflector markers and a not-so-inexpensive Vicon MX motion capture system, which feeds the current position and orientation of the car back to the desktop computer with very high temporal and spatial resolution and low lag (millimeters and milliseconds, respectively). Using this setup, we wrote a simple hard-coded control algorithm that could successfully do point-to-point car racing [34], which is car racing without walls and with randomised waypoints within a bounded area. Even with the excellent sensor data provided by the motion capture system, a physical car has quite a few quirks which are not straightforward to recreate in simulation. These include the response time of the motor and servos, and the variations in acceleration, turning radius and amount of drift depending on such things as what part of the floor the car is on, battery level, motor temperature and whether the car is turning left or right.

But in order to be able to evolve control for the physical car we pretty much need a model: evolving directly on the car would take prohibitively long time and might very well destroy the car in the process. One way of modelling the car would be to measure every dimension, mass, torque, friction and any other relevant detail in the car and in the track, in the attempt of deriving a model from first principles. But this would surely be painstaking, very complex and moreover it would have to be redone every time we changed cars or track. What we want is a way of automatically developing models of vehicle dynamics.

Some first steps toward this was made in a recent paper [35], where we investigated various ways of acquiring models of the *simulated* car based on first person sensor data. We tried to learn neural networks that would take the current sensor inputs at time t and the action taken at this time step, and produce the sensor vector at time $t+1$ as outputs. The results were quite interesting in that they showed that evolution and backpropagation come up with very different types of models (depending on what task you set them, evolution "cheats" in adorably creative but frequently maddening ways), but not very encouraging, in that the models we learned were probably not good enough for forming the basis of a simulator in which we could evolve controllers that could then transfer back to the system that was modelled. We think this is because we are not making any assumptions at all about the sort of function we wanted to learn, and that we could do better by explicitly introducing the basic laws of physics. The approach we are currently investigating is something we call "physics-augmented machine learning": we evolve the parameters of a physics simulation together with a neural network, the idea being that the physics engine provides basic dynamics which are then modulated by the neural network.

But if we do manage to acquire a good model of the vehicle dynamics, could we not then use it as part of the control algorithm as well? In another recent paper [36] we compared evolution with another common reinforcement learning algorithm, temporal difference learning or td-learning. While td-learning can solve some of the same problems evolutionary computation can (but not all), it progresses in quite a different manner, in that it learns during the lifetime of the controller based on local reinforcements and not only between generations, as evolution does. In our experiments, we investigated evolution and td-learning on the point-to-point racing task with and without access to models. Generally, we found evolution to be much more reliable, robust and able to solve more variations of the problem than td-learning, but where td-learning did work it was blazingly fast and could outperform the speed of evolution by orders of magnitude. Clearly, some algorithm that combines the strengths of these two techniques would be most welcome. The comparison of learning control with and without access to a model was less ambiguous: both evolution and td-learning benefited greatly from being able to use the model to predict consequences of their actions, at least in terms of ultimate fitness of the controller. These results go well together with current thinking in embodied cognitive science, which emphasizes the role of internal simulation of action and perception in complex movements and thinking ([37]).

But we must not forget all the environments and models already there in the form of existing commercial racing games. It would be most interesting to see whether the player modelling and track evolution process would work with all the complexities and constraints of a real racing game, and using procedural techniques to generate the visual aspects of the racing track and surrounding environment would be an interesting research topic.

Finally, as controllers grow more sophisticated, we will have to look to more complex vision-like sensor representations and techniques for evolving more complex neural networks. And we will have to keep looking at findings from evolutionary robotics research to integrate into our research on computational intelligence in racing games, as well as trying to export our own findings back into that research community.

10 Resources

Currently, there is no textbook dedicated to computational intelligence in games. If you want to know more about the use of computational intelligence in games in general, a good place to start is either an edited book such as the one by Baba and Jain [38], or the proceedings of the annual IEEE Symposium on Computational Intelligence and games, which showcase the latest work by key researchers in the field [39, 40]. For the application of computational intelligence to racing games in particular, we don't know of any authoritative overview except for this very chapter. Most of the known prior research is referenced in section 2. However, research is progressing fast with several research groups involved, and by the time you read this, new papers are probably out. We are in the process of setting up a web-based bibliography with approximately the same scope as this chapter; this should be reachable through typing the first author's name into a standard Internet search engine.

11 Acknowledgements

We thank Owen Holland, Richard Newcombe and Hugo Marques for their valuable input at various stages of the work described here.

References

1. Fogel, D.B.: Blondie24: playing at the edge of AI. Morgan Kaufmann Publishers Inc., San Francisco, CA, USA (2002)
2. Schraudolph, N.N., Dayan, P., Sejnowski, T.J.: Temporal difference learning of position evaluation in the game of go. In Cowan, J.D., Tesauro, G., Alspector, J., eds.: Advances in Neural Information Processing 6. Morgan Kaufmann, San Francisco (1994) 817–824
3. Runarsson, T.P., Lucas, S.M.: Co-evolution versus self-play temporal difference learning for acquiring position evaluation in small-board go. IEEE Transactions on Evolutionary Computation (2005) 628–640
4. Lucas, S.: Evolving a neural network location evaluator to play ms. pac-man. In: Proceedings of the IEEE Symposium on Computational Intelligence and Games. (2005) 203–210
5. Parker, M., Parker, G.B.: The evolution of multi-layer neural networks for the control of xpilot agents. In: Proceedings of the IEEE Symposium on Computational Intelligence and Games. (2007)
6. Togelius, J., Lucas, S.M.: Forcing neurocontrollers to exploit sensory symmetry through hard-wired modularity in the game of cellz. In: Proceedings of the IEEE 2005 Symposium on Computational Intelligence and Games CIG05. (2005) 37–43
7. Cole, N., Louis, S.J., Miles, C.: Using a genetic algorithm to tune first-person shooter bots. In: Proceedings of the IEEE Congress on Evolutionary Computation. (2004) 13945
8. Spronck, P.: Adaptive Game AI. PhD thesis, University of Maastricht (2005)
9. Miles, C., Louis, S.J.: Towards the co-evolution of influence map tree based strategy game players. In: Proceedings of the IEEE Symposium on Computational Intelligence and Games. (2006)
10. Cliff, D.: Computational neuroethology: a provisional manifesto. In: Proceedings of the first international conference on simulation of adaptive behavior on From animals to animats. (1991) 29–39

11. Stanley, K.O., Bryant, B.D., Miikkulainen, R.: Real-time neuroevolution in the nero video game. IEEE Transactions on Evolutionary Computation **9**(6) (2005) 653–668
12. Nolfi, S., Floreano, D.: Evolutionary robotics. MIT Press, Cambridge, MA (2000)
13. Ampatzis, C., Tuci, E., Trianni, V., Dorigo, M.: Evolution of signalling in a group of robots controlled by dynamic neural networks. In Sahin, E., Spears, W., Winfield, A., eds.: Swarm Robotics Workshop (SAB06). Lecture Notes in Computer Science (2006)
14. Baldassarre, G., Parisi, D., S., N.: Distributed coordination of simulated robots based on self-organisation. Artificial Life **12**(3) (Summer 2006) 289–311
15. Parker, A.: In the blink of an eye. Gardner books (2003)
16. Koster, R.: A theory of fun for game design. Paraglyph press (2004)
17. Tanev, I., Joachimczak, M., Hemmi, H., Shimohara, K.: Evolution of the driving styles of anticipatory agent remotely operating a scaled model of racing car. In: Proceedings of the 2005 IEEE Congress on Evolutionary Computation (CEC-2005). (2005) 1891–1898
18. Chaperot, B., Fyfe, C.: Improving artificial intelligence in a motocross game. In: IEEE Symposium on Computational Intelligence and Games. (2006)
19. Togelius, J., Lucas, S.M.: Evolving controllers for simulated car racing. In: Proceedings of the Congress on Evolutionary Computation. (2005)
20. Togelius, J., Lucas, S.M.: Evolving robust and specialized car racing skills. In: Proceedings of the IEEE Congress on Evolutionary Computation. (2006)
21. Wloch, K., Bentley, P.J.: Optimising the performance of a formula one car using a genetic algorithm. In: Proceedings of Eighth International Conference on Parallel Problem Solving From Nature. (2004) 702–711
22. Stanley, K.O., Kohl, N., Sherony, R., Miikkulainen, R.: Neuroevolution of an automobile crash warning system. In: Proceedings of the Genetic and Evolutionary Computation Conference (GECCO-2005). (2005)
23. Floreano, D., Kato, T., Marocco, D., Sauser, E.: Coevolution of active vision and feature selection. Biological Cybernetics **90** (2004) 218–228
24. Togelius, J., Lucas, S.M.: Arms races and car races. In: Proceedings of Parallel Problem Solving from Nature, Springer (2006)
25. Togelius, J., De Nardi, R., Lucas, S.M.: Making racing fun through player modeling and track evolution. In: Proceedings of the SAB'06 Workshop on Adaptive Approaches for Optimizing Player Satisfaction in Computer and Physical Games. (2006)
26. Togelius, J., De Nardi, R., Lucas, S.M.: Towards automatic personalised content creation in racing games. In: Proceedings of the IEEE Symposium on Computational Intelligence and Games. (2007)
27. Pomerleau, D.A.: Neural network vision for robot driving. In: The Handbook of Brain Theory and Neural Networks. (1995)
28. Bourg, D.M.: Physics for Game Developers. O'Reilly (2002)
29. Monster, M.: Car physics for games. http://home.planet.nl/ monstrous/tutcar.html (2003)
30. Arkin, R.: Behavior-based robotics. The MIT Press (1998)
31. Togelius, J.: Evolved car racing videos. http://video.google.co.uk/videoplay?docid=-3808124536098653151, http://video.google.co.uk/videoplay?docid=2721348410825414373, http://video.google.co.uk/videoplay?docid=-3033926048529222727 (2006)

32. Jirenhed, D.A., Hesslow, G., Ziemke, T.: Exploring internal simulation of perception in mobile robots. In: Proceedings of the Fourth European Workshop on Advanced Mobile Robots. (2001) 107–113
33. Yannakakis, G.N., Maragoudakis, M.: Player modeling impact on player entertainment in computer games. In: User Modeling. (2005) 74–78
34. Togelius, J., De Nardi, R.: Physical car racing video. http://video.google.co.uk/videoplay?docid=-2729700536332417204 (2006)
35. Marques, H., Togelius, J., Kogutowska, M., Holland, O.E., Lucas, S.M.: Sensorless but not senseless: Prediction in evolutionary car racing. In: Proceedings of the IEEE Symposium on Artificial Life. (2007)
36. Lucas, S.M., Togelius, J.: Point-to-point car racing: an initial study of evolution versus temporal difference learning. In: Proceedings of the IEEE Symposium on Computational Intelligence and Games. (2007)
37. Holland, O.E.: Machine Consciousness. Imprint Academic (2003)
38. Baba, N., Jain, L.C.: Computational Intelligence in Games. Springer (2001)
39. Kendall, G., Lucas, S.M.: Proceedings of the IEEE Symposium on Computational Intelligence and Games. IEEE Press (2005)
40. Louis, S.J., Kendall, G.: Proceedings of the IEEE Symposium on Computational Intelligence and Games. IEEE Press (2006)

Evolutionary Algorithms for Board Game Players with Domain Knowledge

Kyung-Joong Kim and Sung-Bae Cho

Department of Computer Science, Yonsei University,
134 Shinchon-dong, Sudaemoon-Ku, Seoul 120-749, South Korea
{kjkim,sbcho}@cs.yonsei.ac.kr

Abstract. Incorporating a priori knowledge, such as expert knowledge, meta-heuristics, human preferences, and most importantly domain knowledge discovered during evolutionary search, into evolutionary algorithms has gained increasing interest in recent years. In this chapter, we present a method for systematically inserting expert knowledge into evolutionary board game framework at the opening, middle, and endgame stages. In the opening stage, openings defined by the experts are used. In this work, we use speciation techniques to search for diverse strategies that embody different styles of game play and combine them using voting for higher performance. This idea comes from the common knowledge that the combination of diverse well-playing strategies can defeat the best one because they can complement each other for higher performance. Finally, we use an endgame database. Experimental results on checkers and Othello games show that the proposed method is promising to evolve better strategies.

1 Introduction

There are two extreme approaches to develop game strategies. One is to use expert domain knowledge extensively in opening, middle and endgame stages. For example, Chinook, the best checkers player in the world, used huge opening and endgame databases with the expert's comments on feature selection and weight determination of evaluation function [1]. It shows very successful to get the world championship-level performance but they need much effort to do that. The other is to use an evolutionary computation to extract novel and good strategies without expert knowledge. The procedure of strategy induction is based on the competition among randomly generated individuals and genetic operations for guiding strategy search. For example, Fogel evolved master-level checkers players from the pure evolution [2].

Although both of them provide a way to create game strategies, they have a limitation to be a more practical one. To solve the problem, hybrid of both approaches was proposed and realized in many different games [3, 4, 5].

K.-J. Kim and S.-B. Cho: *Evolutionary Algorithms for Board Game Players with Domain Knowledge*, Studies in Computational Intelligence (SCI) **71**, 71–89 (2007)
www.springerlink.com © Springer-Verlag Berlin Heidelberg 2007

The focus of the approach is to utilize expert knowledge in evolutionary algorithms. The domain knowledge can be exploited in many parts of evolutionary algorithms; genetic operators, selection, and competition. The incorporation of domain knowledge guides the search direction of evolutionary algorithms or reduces the search spaces and improves the performance of search. Although the incorporated knowledge is not complete compared to the knowledge-based approach, the evolutionary algorithm can be enhanced by the information.

Many board games including Othello, Go, Chess, and Checkers have similar properties. They are perfect information of games and there is no hidden information. Usually, each board game has three different stages: opening, middle, and endgame stages. In opening stage, each game has its own book moves. Expert players are very good at memorizing or organizing a huge number of opening moves. A small disadvantage in an early stage of game results in a great loss after the opening. In the middle stage, it is difficult to define the helpful knowledge explicitly. At the endgame stage, there are relatively small possible choices and it allows programs to calculate the win/draw/loss of the game perfectly. The endgame positions are memorized in a form of DB or perfectly solved in a few seconds.

Opening and endgame knowledge of board games are easily accessible through Internet though they are not enough to make world-level champion program. Board game associations of each country, world organization of board games and many basic introductory materials (books and internet website) provide well-defined knowledge about the games. Using the knowledge, an evolutionary induction approach for board games can be enhanced. Though evolutionary algorithm gives much freedom to the designer, it needs a long evolution time and sometimes, very simple tactics are not evolved or ignored. The idea of knowledge incorporation to the board game is summarized in the figure 1.

In this chapter, the representative two board games, Othello and Checkers, are used to illustrate the feasibility of the approach. Both of them are very well-known games that are frequently used in AI community. Opening and endgame knowledge are well-defined for the games. They are involved in the evolution of board evaluator represented neural networks and weights

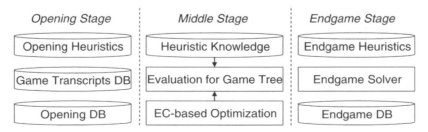

Fig. 1. Conceptual diagram of knowledge-incorporated evolutionary induction of strategies (EC = Evolutionary Computation)

matrix. In Checkers, neural network evaluator is evolved with opening and endgame knowledge. In Othello, weight piece counter is evolved with opening knowledge.

2 Related Works

2.1 Traditional Game Programming

Usually a game can be divided into three general phases: the opening, the middle game, and the endgame. Entering thousands of positions in published books into the program is a way of creating an opening book. The checkers program Colossus has a book of over 34,000 positions that were entered manually [6]. A problem with this approach is that the program will follow published play, which is usually familiar to the humans [7]. Without using an opening book, some programs find many interesting opening moves that stymie a human quickly. However, they can also produce fatal mistakes and enter a losing configuration quickly because a deeper search would have been necessary to avoid the mistake. Humans have an advantage over computers in the opening stage because it is difficult to quantify the relevance of the board configuration at an early stage. To be more competitive, an opening book can be very helpful but a huge opening book can make the program inflexible and without novelty.

The one of important parts of game programming is to design the evaluation function for the middle stage of the game. The evaluation function is often a linear combination of features based on human knowledge, such as the number of kings, the number of checkers, the piece differential between two players, and pattern-based features. Determining components and weighting them require expert knowledge and a long trial-and-error tuning. Attempts have been made to tune the weights of the evaluation function through automated processes, by using linear equations, neural networks, and evolutionary algorithms and can compete with hand-tuning [5, 8].

In chess, the final outcome of most games is usually decided before the endgame and the impact of a prepared endgame database is not particularly significant. In Othello, the results of the game can be calculated in real-time if the number of empty spaces is less than 26. In these two games, the necessity of the endgame database is very low, but in checkers the usage of an endgame database is extremely beneficial. Chinook has perfect information for all checkers positions involving 8 or fewer pieces on the board, a total of 443,748,401,247 positions. These databases are now available for download. The total download size is almost 2.7GB (compressed) [9]. Recently, the construction of a 10-piece database has been completed.

2.2 Evolving Game Players

Checkers is the board game for which evolutionary computation has been used to evolve strategies. Fogel et al. have explored the potential for a co-evolutionary process to learn how to play checkers without relying on the usual inclusion of human expertise in the form of features that are believed to be important to play well [10, 11, 12, 13]. After only a little more than 800 generations, the evolutionary process generated a neural network that can play checkers at the expert level as designated by the U.S. Chess Federation rating system. In a series of six games with a commercially available software product named "Hoyle's Classic Games," the neural network scored a perfect six wins [14]. A series of ten games against a "novice-level" version of Chinook, a high-level expert, resulted in 2 wins, 4 losses, and 4 draws [15].

Othello is a well-known and challenging game for human players. Chong et al. applied Fogel's checkers model to Othello and reported the emergence of mobility strategies [16]. Wu et al. used fuzzy membership functions to characterize different stages (opening game, mid-game, and end-play) in the game of Othello and the corresponding static evaluation function for each stage was evolved using a genetic algorithm [17]. Moriarty et al. designed an evolutionary neural network that output the quality of each possible move at the current board configuration [19]. Moriarty et al. evolved neural networks to constrain minimax search in the game of Othello [20, 21]. At each level, the network saw the updated board and the rank of each move and only a subset of these moves were explored.

The SANE (symbiotic, adaptive neuro-evolution) method was used to evolve neural networks to play the game of Go on small boards with no preprogrammed knowledge [22]. Stanley et al. evolved a roving eye neural network for Go to scale up by learning on incrementally larger boards, each time building on knowledge acquired on the prior board [23]. Because Go is very difficult to deal with, they used a small size board, such as 7×7, 8×8, or 9×9. Lubberts et al. applied competitive co-evolutionary techniques of competitive fitness sharing, shared sampling, and a hall of fame to the SANE neuro-evolution method [24]. Santiago et al. applied an enforced subpopulation variant of SANE to Go and an alternate network architecture featuring subnetworks specialized for certain board regions [25].

Barone et al. used evolutionary algorithms to learn to play games of imperfect information – in particular, the game of poker [26]. They identified several important principles of poker play and used them as the basis for a hypercube of evolving populations. Co-evolutionary learning was used in backgammon [27] and chess [5, 28]. Kendall et al. utilized three neural networks (one for splitting, one for doubling down, and one for standing/hitting) to evolve blackjack strategies [29]. Fogel reported the experimental results of evolving blackjack strategies that were performed about 17 years ago in order to provide some baseline for comparison and inspiration for future research [30]. Ono et al. utilized co-evolution of artificial neural networks on a game

Table 1. Summarization of related-works in evolutionary games

Information	Game	Reference	Knowledge Incorporation
Perfect Information	Checkers	2	
		16	
	Othello	17	O
		19	
		20	
	Go	22	
	Chess	28	O
		5	O
	Kalah	31	O
Imperfect Information	Backgammon	27	
	Poker	26	O
	Blackjack	29	O
		30	O

called Kalah and the technique closely followed the one used by Chellapilla and Fogel to evolve the successful checkers program Anaconda (also known as Blondie24) [31]. Table 1 summarizes the related works. Fogel's checkers framework was used in other games such as [5, 16, 29] and [31]. Fogel et al. applied the framework to the game of chess and reported that the evolved program performed above the master level 5. More information about evolutionary game players is available at [32].

2.3 Othello and Evolution

Because the strategy of Othello is very complex and hard to study even for human, researchers use the game as a platform of AI research. Miikkulainen et al. used evolutionary neural network as a static evaluator of the board [19]. Because they used marker-based encoding for neural network, the architecture and weights were coevolving. In their work, they did not use game tree but the evolution generated mobility strategy, one of the important human strategies for the game. The analysis by the world champion Othello player, David Shaman, was attached and the result showed the possibility of evolutionary approach for the game. In other works, they used evolutionary neural network to focus the search of game tree [20, 21]. In the work, game tree was used to find the good move with static evaluator. The purpose of the evolution was to find a neural network that decided whether deep search was needed. By reducing the moves for deep search, it was possible to expand the search depth more than previous one.

Chong et al. used the same framework used in successful evolutionary checkers to evolve Othello strategies [16, 33]. Their work showed that the same architecture could be successful in another game (Othello). They discussed that the reasons of success were spatial preprocessing layer, self-adaptive mutation, and tournament-style selection (coevolving). They evolved only the weights of fixed-architecture neural network and the concept of spatial preprocessing (dividing the board into a number of sub-boards) were used.

Sun et al. proposed dynamic fitness function for the evolution of weights of linear static evaluator for Othello [18]. The fitness function was changed according to the performance of the previous stage. They also investigated the expansion of chromosome structures to meet new challenges from the environment in the game of Othello [17]. In other work, they used multiple coaches (they were used for fitness evaluation) selected from local computer Othello tournament before evolution [34]. Sun's work was focused on increasing the diversity and self-adaptation of evolutionary algorithm given linear evaluation function. The purpose of evolution was to adjust the weights of features (position, piece advantage, mobility and stability) by expert.

Alliot et al. proposed a method for improving the evolution of game strategies [35]. They evolved weights of linear evaluation function (it was similar to the weight piece counter). The sharing scheme and sophisticated method for fitness evaluation were used. Because Othello's rule was so simple to understand, it was frequently used for educational purpose to teach student about the evolutionary mechanism [36, 37]. Smith et al. proposed co-evolutionary approach for Othello [38]. The fitness of the member in the population was evaluated by the competition results with other members. There were some works about reinforcement learning or self-teaching neural networks for Othello [39, 40]. The discussion about the relationships between reinforcement learning and the evolutionary algorithm can be found at [41].

3 Evolutionary Game Players with Domain Knowledge

3.1 Evolutionary Checkers Framework with Speciation

Usually, an evaluation function is the linear sum of the values of relevant features selected by experts. Input of the evaluation function is the configuration of the board and the output of the function is a value of quality. Designing the function manually needs expertise in the game and tedious trial-and-error tuning. Some features of the board evaluation function can be modeled using machine learning techniques such as automata, neural networks, and Bayesian networks. There are some problems for learning the evaluation function such as determining the architecture of the model and transformation of the configuration into numerical form.

The feed-forward neural network, which has three hidden layers comprising 91 nodes, 40 nodes, and 10 nodes, respectively is used as an evaluator. Multiple neural networks are generated from evolutionary algorithm such as speciation and they are combined to improve the performance. The board configuration is an input to the neural network that evaluates the configuration and produces

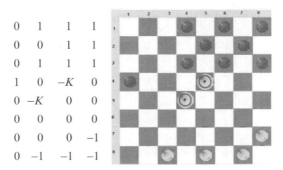

Fig. 2. An example of board representation. Minus sign means the opponent checkers and K means the King. The value of K is evolved with the neural networks

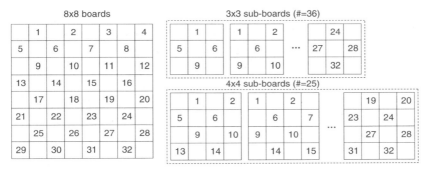

Fig. 3. An example of 4×4, and 3×3 sub-boards. In a checkerboard, there are 91 sub-boards (36 3×3 sub-boards, 25 4×4 sub-boards, 16 5×5 sub-boards, 9 6×6 sub-boards, 4 7×7 sub-boards, and 1 8×8 sub-board). This design gives spatial local information to the neural networks

a score representing the degree of relevance of the board configuration. For evaluation, the information of the board needs to be transformed into the numerical vectors. Each board is represented by a vector of length 32 and components in the vector could have a value of $\{-K, -1, 0, +1, +K\}$, where K is the value assigned for a king, 1 is the value for a regular checker, and 0 represents an empty square. Figure 2 depicts the representation of a board.

To reflect spatial features of the board configuration, sub-boards of the board are used as an input. One board can have 36 3×3 sub-boards, 25 4×4 sub-boards, 16 5×5 sub-boards, 9 6×6 sub-boards, 4 7×7 sub-boards, and 1 8×8 sub-board. 91 sub-boards are used as an input to the feed-forward neural network. Figure 3 shows an example of 3×3, 4×4, and 6×6 sub-boards. The sign of the value indicates whether or not the piece belongs to the player or the opponent. The closer the output of the network is to 1.0, the better the position is. Similarly, the closer the output is to -1.0, the worse the board. Figure 4 shows the architecture of the neural network.

78   K.-J. Kim and S.-B. Cho

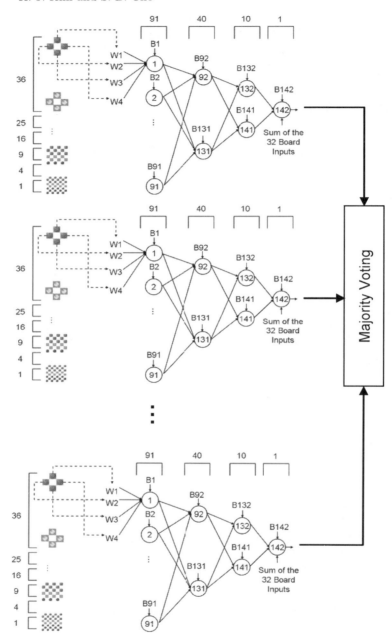

Fig. 4. The architecture of neural network. It is fully connected in the hidden-layers. One sub-board is transformed into the corresponding vector representation and used for the input of the neuron. The output of the neural network indicates the quality of the board configuration

The architecture of the network is fixed and only the weights can be adjusted by evolution. Each individual in the population represents a neural network (weights and biases) that is used to evaluate the quality of the board configuration. Additionally, each neural network has the value of K and self-adaptive parameters for weights and biases. An offspring $P'_i, i = 1, ..., p$ for each parent $P_i, i = 1, ..., p$ is created by

$$\sigma'_i(j) = \sigma_i(j) \exp(\tau N_j(0, 1)), \ j = 1, ..., N_w$$

$$w'_i(j) = w_i(j) + \sigma'_i(j) N_j(0, 1), \ j = 1, ..., N_w$$

where N_W is the number of weights and biases in the neural network (here this is 5046), $\tau = 1/\sqrt{2\sqrt{N_w}} = 0.0839$, and $N_j(0, 1)$ is the standard Gaussian random variable resampled for every j. The offspring king value K' was obtained by

$$K'_i = K_i + \delta$$

where δ was chosen uniformly at random from $\{-0.1, 0, 0.1\}$. For convenience, the value K' was constrained in [1.0, 3.0] by resetting to the limit exceeded when applicable.

In fitness evaluation, each individual chooses five opponents from a population pool and plays games with the players. Fitness increases by 1 for a win while the fitness of an opponent decreases by 2 for a loss. In a draw, the fitness values of both players remain the same. After all the games are played, the fitness values of all players are determined. Deterministic crowding algorithm is used to maintain the diversity of the population. After finishing evolution, the final population is clustered into species and the representative players from each species are combined to play.

3.2 Knowledge Incorporation into Evolutionary Checkers

As mentioned before, we can classify a single checkers game into three stages: opening, middle, and endgame stages. In the opening stage, about 80 previously summarized openings are used to determine the initial moves. In the middle stage, a game tree is used to search for an optimal move and an evolutionary neural network is used to evaluate leaf nodes of the tree. In the neural network community, it is widely accepted that the combination of multiple diverse neural networks can outperform the single network [42]. Because the fitness landscape of an evolutionary game evaluation function is highly dynamic, a speciation technique like fitness sharing is not appropriate. A crowding algorithm that can cope with a dynamic landscape is adopted to generate more diverse neural network evaluation functions. The performance of evolutionary neural networks for creating a checkers evaluation function has been demonstrated by many researchers 2. In the end game, an endgame database is used which indicates the result of the game (win/loss/draw) if the number of remaining pieces is smaller than a predefined number (usually from 2 to 10).

Opening Stage

The opening stage is the most important opportunity to defeat an expert player because trivial mistakes in the opening can lead to an early loss. The first move in checkers is played by red and there are seven choices (9-13, 9-14, 10-14, 10-15, 11-15, 11-16, and 12-16). These numbers refer the labels on the board (the top-left square is 1) and the X-Y means red moves a piece from position X to position Y. Usually, 11-15 is the best move for red but there are many other alternatives. They are described with specific names, such as Edinburgh, Double Corner, Denny, Kelso, Old Faithful, Bristol, and Dundee, respectively. For each choice, there are many well-established additional sequences which range in length from 2 to 10. The longest sequence is described as the White Doctor: 11-16, 22-18, 10-14, 25-22, 8-11, 24-20, 16-19, 23-16, 14-23, 26-19. Careful analysis over decades of tournament play has proven the usefulness or fairness of the opening sequences. Initial sequences are decided by the opening book until the move is out of the book. Each player chooses its opening randomly and the seven first choices have the same probability to be selected as an opening.

If the first move is 9-13 (Edinburgh), there are 8 openings that start from 9-13. They are Dreaded Edinburgh (9-13, 22-18, 6-9), Edinburgh Single (9-13, 22-18, 11-15), The Garter Snake (9-13, 23-19, 10-15), The Henderson (9-13, 22-18, 10-15), The Inferno (9-13, 22-18, 10-14), The Twilight Zone (9-13, 24-20, 11-16), The Wilderness (9-13, 22-18, 11-16), and The Wilderness II (9-13, 23-18, 11-16). In this case, there are four choices for the second moves: 22-18, 23-19, 24-20, and 23-18. The second move is chosen randomly and the next moves are selected continually in the same manner. Table 2 shows an example of openings in checkers.

Table 2. Openings in Checkers game

Name	Opening	Name	Opening
Edinburgh	9-13	Old Faithful	11-15
Dreaded Edinburgh	9-13, 22-18, 6-9	Cross	11-15, 23-18, 8-11
Edinburgh Single	9-13, 22-18, 11-15	Dyke	11-15, 22-17, 15-19
The Garter Snake	9-13, 23-19, 10-15	Bristol	11-16
The Henderson	9-13, 22-18, 10-15	Bristol Cross	11-16, 23-18, 16-20
Double Corner	9-14	Millbury	11-16, 22-18, 8-11
Double Corner Dyke	9-14, 22-17, 11-15, 25-22, 15-19	Oliver's Twister	11-16, 21-17, 8-11
Denny	10-14	Dundee	12-16
Kelso Cross	10-15, 23-18	Bonnie Dundee	12-16, 24-20, 8-12
The Tyne	10-15, 21-17, 9-13	The Skunk	12-16, 24-20, 10-15

Endgame Stage

The estimated quality of the board is calculated using the evolved neural networks to evaluate the leaf nodes of the tree with the minimax algorithm. If the value of f (estimated quality of the next moves) is not reliable, we refer to domain-specific knowledge and revise f. The decision rule for querying the domain knowledge is defined as follows.

IF ($f < 0.75$ and $f > 0.25$) or ($f < -0.25$ and $f > -0.75$)
     THEN querying the domain knowledge

For example, there is a 2-ply game tree and the concept of selective domain-specific knowledge is like this. In the game tree, let's assume that there are 8 terminals. The two choices are evaluated as 0.3 and -0.9. It is clear that the board configuration as evaluated -0.9 is not good. However, the board configuration with 0.3 is not enough to decide as a draw and needs querying the domain knowledge. If the returned answer from the DB is a draw, the player must select the move. However, if the answer is a loss, the player could select the configuration of -0.9.

3.3 Othello

As mentioned before, we have classified a single Othello game into three stages: opening, middle, and endgame stages. In the opening stage, about 99 previously summarized openings are used to determine the initial moves. In the middle and endgame stage, a weight piece counter is used to search for a good move and an evolutionary algorithm is used to optimize the weights of the counter. Though game tree is useful to find a good move, in this paper we only focus on the goodness of the evaluation function. Figure 5 shows the procedural flow of the proposed method in a game. The longest opening has 18 moves and it can significantly reduce the difficulty of the evaluation function optimization.

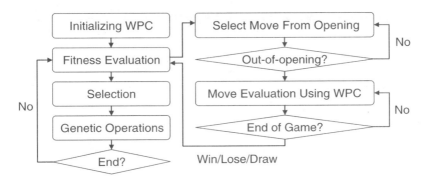

Fig. 5. The flow of the game using the proposed method

Opening Stage

Opening is very important for Othello game because the game is extremely short compared to other games and the importance of early moves is huge. In the games between expert players, gaining small advantage in the early games is very important. They attempt to memorize good or tricky moves before tournament. Brian Rose, World Othello Champion at 2002, said that he memorized about 300 lines before the championships. It is critical to prepare and memorize well-defined openings in the Othello game.

In this work, well-defined 99 openings are used (http://www.oth ello.nl/ content/anim/openings.txt). The length of opening ranges from 2 to 18. In fact, in the expert player's game, there is a case that all moves are memorized by both of them. The length of opening can be expanded to the last move, but it is not possible to exploit all of them. "Out-of-opening" means that the moves by the player are not in the list of the openings. Because it is carefully analyzed by experts, it can guarantee the equivalence between two players. In some openings, they are a bit beneficial to a specific color but it is not important because in human's game the difference can be recovered in the middle and end game.

The opening is named after their shapes. Tiger, Cat, and Chimney means that the shape of the opening is similar to the objects. Until out-of-opening, the WPC player follows one of the openings in the list. Figure 6 shows an example of opening selection. The current sequence of the game is defined as $S = \{m_1, m_2, ..., m_n\}$. Here, m_i represents each move, and n is the number of moves played. For each opening, we check whether the first n moves are the same with the sequence. The satisfactory openings are called candidates.

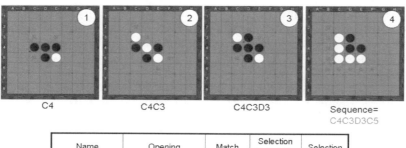

Fig. 6. The selection of opening (Cow is selected)

Among candidates, one opening is chosen probabilistically. The next move of the selected opening after the first n moves is decided as the next move of player. At the 3rd move, there are 3 choices: Diagonal, perpendicular, parallel. For each choice, there are several extended openings. The probability of choice for each line is determined based on the number of openings. For example, the diagonal opening has 59 variations in his line and about 59% probability to be selected. If there are many variations for each choice, it means that the line is the most preferable one for humans. The probability of selection is based on the human's long investigation.

The probability of opening move selection can be determined using another way. For example, the probability of openings in the games of WTHOR database (It contains all public games played between humans) can be used. Otherwise, specific book evaluation (automatically learned from self-played games of strong program) values can be exploited. WZEBRA (strong Othello program) provides evaluation value for each opening. For example, the X-square opening is -23 (in black's perspective) and Snake is -4.

Evolving Evaluator (WPC)

The evaluation of each WPC (Weighted Piece Counter) is based on the competition against static player. Standard heuristic WPC is used to evaluate the performance of the individual. Figure 7 depicts the standard WPC. Its darkness represents the weight of each position. In Othello, four corners are extremely important and the weight of the corners is 1.0. Other places except the ones near the corner has the similar (0.01~0.1) weights. The positions near the corners are very dangerous because it gives a chance to the opponent to capture the corner. Because it is static, it cannot evaluate well the dynamics of the relevance of position, but it is still strong compared to other random approaches.

The fitness of the WPC is evaluated based on the 1000 games between the standard WPCs. Given two deterministic strategies, there can be only two games (changing the color). This makes the fitness evaluation difficult and

1.0	−0.25	0.1	0.05	0.05	0.1	−0.25	1.0
0.25	−0.25	0.01	0.01	0.01	0.01	−0.25	−0.25
0.1	0.01	0.05	0.02	0.02	0.05	0.01	0.1
0.05	0.01	0.02	0.01	0.01	0.02	0.01	0.05
0.05	0.01	0.02	0.01	0.01	0.02	0.01	0.05
0.1	0.01	0.05	0.02	0.02	0.05	0.01	0.1
0.25	−0.25	0.01	0.01	0.01	0.01	−0.25	−0.25
1.0	−0.25	0.1	0.05	0.05	0.1	−0.25	1.0

Fig. 7. Weights of the heuristic

random moves are used. 10% of moves of both players are decided randomly. It allows 1000 games among them being different but it also makes the fitness of each WPC unstable. To preserve good solutions for the next generation, elitist approach is used.

If the number of wins is W, the number of loses is L, and the number of draws is D, the fitness is defined as follows.

$$fitness = W * 1.0 + D * 0.5$$

The weights are widely used in real Othello tournament.

The weights of each position are initialized as a value between -1 to 1. Until out-of-opening, the WPC in the population uses opening knowledge. After out-of-opening, the WPC evaluates the relevance of board and decides the next move. Among possible moves, the one with the highest WPC is selected. Roulette-wheel selection is used and 1-point crossover is applied to the converted 1-dimensonal array of the 8×8 board. Mutation operator changes an element of the vector as a new value ranging from -1 to 1.

4 Experimental Results

4.1 Checkers

Speciation algorithm is used to generate diverse strategies and improve the performance of evolutionary strategies by combining multiple strategies [43]. The non-speciated evolutionary algorithm uses a population size of 100 and limits the run to 50 generations. The speciated evolutionary algorithm sets the population size to 100, generations to 50, the mutation rate to 0.01 and crossover rate to 1.0. The number of leagues (used to select the best player from each species) is 5 (5 means that each player selects 5 players from the species randomly and the competition results are used for the selection). Evolving checkers using speciation requires 10 hours on a Pentium III 800MHz (256MB RAM). The non-speciated evolutionary algorithm uses only mutation but the speciated evolutionary algorithm uses crossover and mutation. The non-speciated evolutionary algorithm is the same as Chellapilla and Fogel's checkers program. The parameters of a simple EA are Population_Size=100, Mutation_rate=1.0, Generation=50. The parameters of the speciated EA are Population_Size=100, Mutation_rate=0.01, Crossover_rate=1.0 and Generation=50. They have the same number of individuals for evolution and the game that they played for one generation is the same to ours.

The Chinook endgame database (2~6 pieces) is used for revision when the estimated value from the neural network is between 0.25 and 0.75 or between -0.25 and -0.75. Time analysis indicates that the evolution with knowledge takes much less time than that without knowledge in simple evolution (Figure 8) and the knowledge-based evolution takes a little more time than that without knowledge in the speciated evolution (Figure 8). This means that

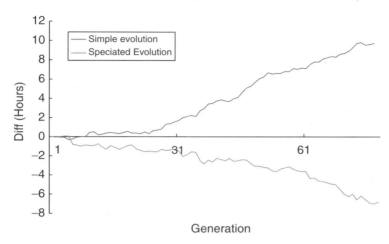

Fig. 8. Comparison of running time (Diff = Time(Without Knowledge)-Time(With Knowledge))

Table 3. The competition results between the speciated players using both opening and endgame DB and the speciated player with one of the knowledge (Win/Lose/Draw)

Speciated EA with opening and endgame DB	Speciated EA with opening DB	14/6/10
Speciated EA with opening and endgame DB	Speciated EA with endgame DB	9/10/11
Speciated EA with opening DB	Speciated EA with endgame DB	5/13/12

the insertion of knowledge within a limited scope can accelerate the speed of the evolutionary algorithm because it can reduce the computational requirement for finding an optimal endgame sequence by using the endgame DB. Since we have used two different machines to evolve simple and speciated versions, respectively, direct comparison of evolution time between them is meaningless. In the speciated evolution, the insertion of knowledge increases the evolution time and an additional 5 hours are needed for 80 generations. Table 3 shows the effect of the stored knowledge (opening and endgame databases) in speciation.

4.2 Othello

The parameters of the evolution are as follows. The maximum generation is 100, population size is 50, crossover rate is 0.8, and mutation rate is 0.01. The best fitness is about 540 in the evolution of WPC with opening knowledge. After then, it converges to 500 though the population size is 50. The best individual at the 1st generation is 400. It means that the incorporation of the

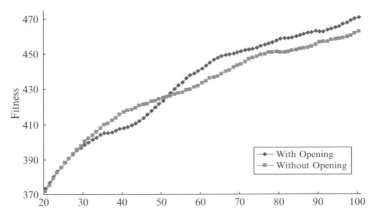

Fig. 9. Average of 10 runs. The value is average of 20 generations to show trends

opening knowledge gives some individuals the high performance gain. The average fitness at the 1st generation is about 200. It is not quite different with the one of evolution without opening. That means the incorporation of opening does not increase the performance of all individuals in the population.

The evolution finds solutions that can do well after the opening and it can significantly reduce the burden for the WPC. Though the generations from 21 to 45 show performance degradation, the best individual that exploits the advantage of opening emerges at 51. The comparison of the maximum fitness over the evolution between the one with opening and the one without opening shows that the one with opening shows better fitness during evolution. Though the competition dose not support opening and the direct comparison against them is not possible, the 550 is about rank 12 in the CEC competition in 2006 [44].

The comparison of the average fitness between them shows that the average fitness of the one with opening is larger than that without opening, but it finally converges to the same value. It means the incorporation of opening does not give huge advantage to the population. The average fitness of the 1st generation is not different and it means that the evolution finds more improved individuals that exploit the opening knowledge. Figure 9 shows the average of 10 runs.

5 Concluding Remarks

In this chapter, we presented several methods to incorporate domain knowledge into evolutionary board game players: checkers and Othello. Well-defined opening and endgame knowledge can be easily incorporated into the evolutionary induction procedure of strategies and experimental results confirm that the presented methods can improve the performance of evolved strategies compared to that without knowledge.

Acknowledgements

This research was supported by Brain Science and Engineering Research Program sponsored by Korean Ministry of Commerce, Industry and Energy.

References

1. Schaeffer, J.: One Jump Ahead: Challenging Human Supremacy in Checkers. Springer (1997)
2. Fogel, D.B.: Blondie 24: Playing at the edge of AI. Morgan Kaufmann (2001)
3. Kim, K.-J. and Cho, S.-B.: Systematically incorporating domain-specific knowledge into evolutionary speciated checkers players. IEEE Transactions on Evolutionary Computation, 9(6) (2005) 615-627
4. Kim, K.-J. and Cho, S.-B.: Evolutionary Othello players boosted by opening knowledge. World Congress on Computational Intelligence (2006)
5. Fogel, D.B., Hays, T.J., Hahn, S. and Quon, J.: A self-learning evolutionary chess program. Proc. of the IEEE 92(12) (2004) 1947-1954
6. Chernev, I.: The Compleat Draughts Player. Oxford University Press, (1981)
7. Schaeffer, J., Culberson, J., Treloar, N., Knight, B., Lu, P. and Szafron, D.: A world championship caliber checkers program. Artificial Intelligence 53(2-3) (1992) 273-290
8. Schaeffer, J., Lake, R., Lu, P. and Bryant, M.: Chinook: The man-machine world checkers champion. AI Magazine 17(1) (1996) 21-29
9. R. Lake, J. Schaeffer and P. Lu, "Solving large retrograde analysis problems using a network of workstations," *Proc. of Advances in Computer Chess VII*, pp. 135-162, 1994.
10. Fogel, D.B.: Evolutionary entertainment with intelligent agents. IEEE Computer, 36(6) (2003) 106-108
11. Chellapilla, K. and Fogel, D.B.: Evolving neural networks to play checkers without relying on expert knowledge. IEEE Trans. on Neural Networks 10(6) (1999) 1382-1391
12. Chellapilla, K. and Fogel, D.B.: Evolving an expert checkers playing program without using human expertise. IEEE Trans. on Evolutionary Computation 5(4) (2001) 422-428
13. Fogel, D.B.: Evolving a checkers player without relying on human experience. ACM Intelligence, 11(2) (2000) 20-27
14. Chellapilla, K. and Fogel, D.B.: Anaconda defeats Hoyle 6-0: A case study competing an evolved checkers program against commercially available software. Proc. of the 2000 Congress on Evolutionary Computation 2 (2000) 857-863
15. Fogel, D.B. and Chellapilla, K.: Verifying Anaconda's expert rating by competing against Chinook: Experiments in co-evolving a neural checkers player. Neurocomputing 42(1-4) (2002) 69-86

16. Chong, S.Y., Tan, M.K. and White, J.D.: Observing the evolution of neural networks learning to play the game of Othello. IEEE Transactions on Evolutionary Computation 9(3) (2005) 240-251
17. Sun, C.-T. and Wu, M.-D.: Multi-stage genetic algorithm learning in game playing. NAFIPS/IFIS/NASA '94 (1994) 223-227
18. Sun, C.-T. and Wu, M.-D.: Self-adaptive genetic algorithm learning in game playing. IEEE International Conference on Evolutionary Computation 2 (1995) 814-818
19. Moriarty, D.E. and Miikkulainen, R.: Discovering complex Othello strategies through evolutionary neural networks. Connection Science 7 (1995) 195-209
20. Moriarty, D. E. and Miikkulainen, R.: Evolving neural networks to focus minimax search. Proc. of the 12th National Conf. on Artificial Intelligence (AAAI-94) (1994) 1371-1377
21. Moriarty, D.E. and Miikkulainen, R.: Improving game-tree search with evolutionary neural networks. Proc. of the First IEEE Conf. on Evolutionary Computation, 1 (1994) 496-501
22. Richards, N., Moriarty, D. and Miikkulainen, R.: Evolving neural networks to play go. Applied Intelligence 8 (1998) 85-96
23. Stanley, K.O. and Miikkulainen, R.: Evolving a roving eye for go. Proc. of the Genetic and Evolutionary Computation Conference (GECCO-2004) (2004) 1226-1238
24. Lubberts, A. and Miikkulainen, R.: Co-evolving a go-playing neural networks. Proc. of 2001 Genetic and Evolutionary Computation Conference Workshop Program (GECCO-2001), (2001) 14-19
25. Perez-Bergquist, A.S.: Applying ESP and region specialists to neuro-evolution for go. Technical Report CSTR01-24, Department of Computer Science, The University of Texas at Austin, May (2001)
26. Barone, L. and While, L.:An adaptive learning model for simplified poker using evolutionary algorithms. Proc. of the 1999 Congress on Evolutionary Computation 1 (1999) 153-160
27. Pollack, J.B. and Blair, A.D.: Co-evolution in the successful learning of backgammon strategy. Machine Learning 32(3) (1998) 225-240
28. Kendall, G. and Whitwell, G.: An evolutionary approach for the tuning of a chess evaluation function using population dynamics. Proc. of the 2001 Congress on Evolutionary Computation 2 (2001) 995-1002
29. Kendall, G. and Smith, C.: The evolution of blackjack strategies. Proc. of the 2003 Congress on Evolutionary Computation 4 (2003) 2474-2481
30. Fogel, D.B.: Evolving strategies in blackjack. Proc. of the 2004 Congress on Evolutionary Computation (2004) 1427-1434
31. Ono, W. and Lim, Y.-J.: An investigation on piece differential information in co-evolution on games using Kalah. Proc. of Congress on Evolutionary Computation 3 (2003) 1632-1638
32. Lucas, S.M. and Kendall, G.: Evolutionary computation and Games. IEEE Computational Intelligence Magazine (2006) 10-18
33. Chong, S.Y., Ku, D.C., Lim, H.S., Tan, M.K. and White, J.D.: Evolved neural networks learning Othello strategies. The 2003 Congress on Evolutionary Computation 3 (2003) 2222-2229
34. Sun, C.-T., Liao, Y.-H., Lu, J.-Y. and Zheng, F.-M.: Genetic algorithm learning in game playing with multiple coaches. IEEE World Congress on Computational Intelligence 1 (1994) 239-243

35. Alliot, J. and Durand, N.: A genetic algorithm to improve an Othello program. Artificial Evolution (1995) 307-319
36. Eskin, E. and Siegel, E.: Genetic programming applied to Othello: Introducing students to machine learning research. Proceedings of the Thirtieth SIGCSE Technical Symposium on Computer Science Education, 31(1) (1999) 242-246
37. Bateman, R.: Training a multi-layer feedforward neural network to play Othello using the backpropagation algorithm and reinforcement learning. Journal of Computing Sciences in Colleges 19(5) (2004) 296-297
38. Smith, R.E. and Gray, B.: Co-adaptive genetic algorithms: An example in Othello Strategy. TCGA Report No. 94002, The University of Alabama (1994)
39. Leuski, A. and Utgoff, P.E.: What a neural network can learn about Othello. Technical Report TR96-10, Department of Computer Science, University of Massachusetts, Amherst (1996)
40. Leuski, A.: Learning of position evaluation in the game of Othello. Technical Report TR 95-23, Department of Computer Science, University of Massachusetts, Amherst (1995)
41. Moriarty, D.E., Schultz, A.C. and Grefenstette, J.J.: Evolutionary algorithms for reinforcement learning. Journal of Artificial Intelligence Research 11 (1999) 241-276
42. Hansen, L.K., and Salamon, P.: Neural network ensembles. IEEE Trans. on Pattern Analysis and Machine Intelligence 12(10) (1990) 993-1001
43. Kim, K.-J. and Cho, S.-B.: Evolving speciated checker players with crowding algorithm. Congress on Evolutionary Computation (2002) 407-412
44. CEC 2006 Othello Competition, http://algoval.essex.ac.uk:8080/othello/html/Othello.html.

The ChessBrain Project – Massively Distributed Chess Tree Search

Colin Frayn[1] and Carlos Justiniano[2]

[1]Cercia, School of Computer Science, University of Birmingham, UK
[2]Production Technologies, Artificial Intelligence Group, Countrywide Financial Corporation, Agoura Hills, California

1 Introduction

ChessBrain was formed in January 2002 to investigate distributed computing concepts, including distributed game tree search, using a network of volunteer contributors linked together over the Internet. The project brought together a worldwide team of experts to work on the various aspects of the task in parallel. The most important components being the distributed communication protocol, client verification and security components, and the chess-playing engine itself.

Chess lends itself to the concept of distributed computation by virtue of the parallel nature of game-tree search. The paradigm is very simple, and can be explained with a simple metaphor. Imagine sitting in front of a chessboard, ready to play a move against a grandmaster opponent. You have twenty potential moves that you are considering, so you phone twenty chess-playing friends, and ask each one of them to analyze a separate move, and quickly report back with their verdict.

After a while, the first replies arrive, from those easily-analysed moves which were deemed trivial losses (losing material almost instantly with no prospect of recovery.) Somewhat later, other friends return with results from some of the more complicated moves. After some time, you have gathered nineteen of the results, just waiting on one last move. Your twentieth friend phones back and announces that the twentieth move is too complicated, so he suggests that you further subdivide it and phone up your friends once more to get them to analyze the components of this one potential move. Once this move has been divided and analysed, the second set of results comes back, and one of your friends is particularly enthusiastic about a particular line that looks very strong indeed. After comparison with the previous results, this new line seems to be the strongest, so you play the move and stop the clock.

This is parallelized game tree search applied to chess. The algorithm seems very simple indeed, though of course, as with most such projects, the complexity comes from the details. In this chapter we will firstly give an overview of the algorithms required for chess tree search, and the infrastructure and technical components required to get the ChessBrain project working. Secondly, we will cover many of those same details that make this project such a difficult and worthwhile one to attempt. Thirdly, and finally, we discuss the second-generation ChessBrain II project, which solved many of the drawbacks of our initial version and which is being prepared for public release some time within the coming year.

2 Computer Chess

The basic algorithms for game tree search have been covered in detail in the literature, though parallel game tree search is less well studied. We include an introduction to 'how computers play chess', together with an overview of parallel game tree search.

2.1 Game Tree Search

Chess game tree analysis depends on the concept of recursive search. The fundamental proposition is simply that the strength of any one position depends negatively on the strength of the opponent's best move. The stronger the opponent's reply is, the weaker your position.

The first algorithm for chess playing in a computer was devised by British mathematician Alan Turing in the year 1947. At that time, no computers existed on which Turing's program could be executed, though Turing felt sure that computers would soon reach the same standard in chess as human beings. The first article about writing an algorithm to play computer chess was written in 1950 by Claude Shannon [1] which introduced many of the foundational concepts that are still employed today.

The basic algorithm for game tree search in chess is the same used for many simple zero-sum board games. It is a recursive, limited-depth search algorithm called *minimax search* [2], which simply uses the recursive dividing of board positions in order to expand the game tree. In this sense, one examines the strength of any one move by calculating the strength of the opponent's best reply. In order to evaluate the opponent's best reply, one must next examine each of the opponent's replies, in turn, to each of one's own possible moves. In order to evaluate these replies, one must analyse one's own replies to each of them. And so on (potentially) *ad infinitum*.

```
Loop through all moves
      Play move
      move_score = - Opponent's_best_move_score
      if (move_score > best_move_score)
```

```
        then (best_move = move)
     Undo Move
End of Loop
```

Though this algorithm plays theoretically perfect chess, it would take an enormously long time to do so as the game tree thus generated would expand exponentially with a branching factor approaching 35 moves per position, and games lasting at least 60–80 moves in total. This leads to estimates that the total game tree size for chess is approximately 10^{123} nodes, which is impossibly large for standard search techniques. [3]

So conventional chess playing simply comes down to searching as much of the available game tree as possible in the quickest time possible, whilst making as few inaccurate generalizations as possible. Game tree search effectively becomes an error minimization problem. Once one has identified a suitable subset of the total tree, then one evaluates this game tree to a certain number of levels (or 'ply') and then one applies an evaluation function at the leaf nodes which aims to duplicate as accurately as possible the expected future gain from the position reached. Any inaccuracy in computer play is due to the fact that these terminal evaluations do not always accurately reflect the chances of winning from the given positions.

Pseudo code for this algorithm is as follows:

```
(Set depth initially to required value)
SEARCHING_FUNCTION {
     Decrease depth by 1
     Loop through all moves
        Play move
        if (depth = 0)
          move_score = static_position_score
        else move_score = - Opponent's_best_move_score
        if (move_score > best_move_score)
          then (best_move = move)
        Undo Move
     End of Loop
     Return best_move_score
  } END
```

Improving Search Depth

There are three primary ways to improve the depth (and hence the accuracy) of this search technique. Firstly, one can devise increasingly clever pruning algorithms for reducing the game tree size. The most effective of these is the theoretically sound algorithm called *alpha-beta pruning* [4]. This reduces the branching factor to (approximately) its square root, when implemented well, and hence doubles the possible search depth in a fixed time. Alpha-beta pruning works by passing parameters down through the search tree that reflect the window of interest for that section of the search. If, at any point, a search branch scores in excess of the maximum (beta) level then that entire search is aborted back to the previous ply.

A simplified version of the algorithm for alpha-beta search is as follows:

```
initially alpha = -INFINITY, beta=INFINITY
search(position,side,depth,alpha,beta) {
   best_score = -INFINITY
   for each move {
      do_move(position, move)
        if (depth is 0)
           move_score = static_score(position, side)
        else
           move_score = -search(position, opponent side,
                                depth-1, -beta, -alpha)
      undo_move(position,move)
      if (move_score > best_score)
         best_score = move_score
      if (best_score > alpha) alpha = best_score
      if ( alpha >= beta ) return alpha
   }
   return best_score
}
```

Other methods are largely unsound, meaning that they are not guaranteed to retain exactly the same outcome to a search. Most of these involve pruning out branches of the game tree when they appear to be uninteresting, which may be risky unless one actually searches them fully in order to ensure this with 100% accuracy. Often these algorithms reduce unnecessary searching, but occasionally they cause the computer to miss an important variation. An example of this technique is 'razoring', by which a branch of the search tree is ignored if a shallower search indicates that the branch is so disastrous that it is extremely unlikely to end up with a satisfactory result.

A second approach is simply to improve the evaluation function that is used at leaf nodes in the search tree. Clearly, if one can spot an attack one ply earlier then one need search one ply less deeply, thus vastly reducing the size of the game tree. However, this is significantly easier said than done, and although a considerable amount of time and effort is being invested in this area, it is still fair to say that a computer's intuition is significantly weaker than that of a human being. In the game between Deep Blue and Garry Kasparov [5] it was only by searching approximately one hundred million times as many positions per second, that Deep Blue could compete at the same level.

Finally, one can increase the computation that is available to search the game tree, either by purchasing a faster computer, or by distributing the work to a number of computers connected in parallel. Recent competitions against leading grandmasters, for example the Hydra project [6], have focused on creating extremely strong, parallel chess programs that work on powerful, multiprocessor computers. An alternative way to create a large parallel computation network is through volunteer computing. Though there are also numerous drawbacks to this approach, it is this latter model that we will pursue for the remainder of this chapter.

The Horizon Effect

One significant observation that must be made is concerning the so-called *horizon effect*, which affects the leaf nodes of a game tree, where significant changes in the value of a search branch might be 'pushed' just beyond the edge of the search and hence missed. To illustrate the problem, imagine that a shallow search has indicated that capturing a minor piece with your queen is a good move. Without searching one ply deeper, a naïve chess engine might miss the fact that the very next move results in the opponent recapturing the queen, and hence that the branch is very poor.

In order to cope with this problem, chess engines never actually terminate the search at the leaf nodes, but instead they continue the search with a *Quiescence Search*, which continues to search only moves which might have a significant effect on the game balance. That is to say, captures, checks and promotions. Such a search assumes that if none of the non-quiescent moves leaves the side in a better position, then the static leaf score should be returned, as before. However, if there is a capture sequence (for example) that appears strong, then the static evaluation can no longer be accepted, and should be replaced with an estimate of the positional value after such an exchange.

2.2 Parallel Game Tree Search

The distributed search algorithm has already been introduced, albeit anecdotally. As the branches of the game tree do not intersect beneath their branching point, it is safe to distribute the separate branches among available processors in order that they might search them independently and in parallel. This method works well with some simple thought, but once one implements some of the pruning methods mentioned above then the picture becomes more complicated. All the conventional tree search algorithms are based on sequential, shared-memory search where the algorithm investigates each branch in turn, and the time savings (and the most substantial pruning) are dependent on incorporating into the search of each branch, the results of previous searches. Using this method, computers can benefit from past experience, and can update the search targets in light of the results of earlier analysis.

The *alpha-beta* algorithm, mentioned above, is the most important pruning algorithm in chess tree search, and it depends heavily on a serial, shared-memory architecture in order to work perfectly. In order to illustrate this algorithm, consider the following example. Imagine a position in which you have ten possible moves. In order to evaluate each of these moves, you will play the move, and then examine your opponent's potential replies. Your move is as strong as (minus) the opponent's best reply. So if the opponent has a move scoring +X points, then your move scores −X. Chess is a zero-sum game, meaning that every move that is good for one side is bad for the other side by exactly the same amount, because the two sides are in direct opposition.

Suppose we examine our first potential move, where our opponents best reply is valued at only −15 points (from his or her frame of reference), this means that our move scores, from our perspective, +15 points.

Next let's consider the second move. Now we know that we can score *at least* +15 points from this root position, so if this second move is worth less than or equal to +15 points then it is worthless to us. But what does that mean? Well it means that this move is only interesting if our opponent has no moves scoring (for him) ≥ −15 points. If the opponent has just *one* move that scores −15 or better (for him) then he will play it, which means that the *best* our move can score is minus this amount, or +15 points.

So if we search through the opponent's best moves and discover a move that scores (for him) −12 points, we know that we can immediately exit from this search without searching any of the opponent's remaining moves. Why is that? Well simply because we *know* that they will have no effect whatsoever on our choice of move – our opponent can score *at least* −12 points from this line, which means that we can score *at most* +12 points by playing this move, which is less than our best-so-far of +15.

The implications of implementing the *alpha-beta* algorithm within a parallel chess environment are quite serious, because we have no sharing of information between the nodes. When we start searching our second potential move in parallel, we don't yet have the results from searching the first move, so we cannot use the score from the first move to prune the search of the second. This means that we lose a great deal of efficiency in parallel search compared to serial search. The expectation is that we can compensate for this by using many more computers.

Another drawback of fully distributed parallel search is that we lose the ability to share memory between contributors. When searching on one machine, we can construct a *transposition table*, which contains the results of previous searches in case they may be of use at a later time. However, we cannot do this in parallel search because these hash tables generally take up

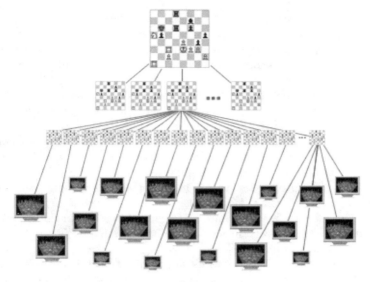

Fig. 1. Distributed chess tree search

tens or hundreds of megabytes of storage space and are updated tens or hundreds of thousands of times per second, which simply cannot be mirrored in real time over thousands of separate machines.

The ChessBrain project's core distributed search uses the APHID algorithm [7] at its heart, which enables a flexible, periodic update of new information as and when it arrives into the existing search tree. It implements an incremental, *iterative deepening* search, firstly locally on the server (known as the 'SuperNode') and then, after a certain fixed time, within the distribution loop. During this latter phase, the top few plies of the search tree are analysed repeatedly with new leaf nodes being distributed for analysis to individual contributors (called 'PeerNodes') as soon as they arise. Information received from the distributed network is then incorporated into the search tree, with branches immediately extended or pruned as necessary.

When a new position is initialised, the SuperNode analyses it locally for a configurable maximum depth or search time. With ChessBrain, it proved most profitable to continue searching for as many plies as possible, but to move to the distributed search component as soon as the current ply was finished and at least ten seconds had elapsed. This threshold was used for two reasons. Firstly, it was not beneficial to use the parallel search with a very shallow ply depth and secondly, the extra information gained during a reasonably thorough 7 or 8 ply local search was extremely useful in the initial move ordering for the distributed game tree. However, when the total time limit for a move was less than a minute, parallel search was not used at all and search proceeded entirely on the SuperNode.

Once the local search was completed, then the parallel search began. In this phase, the SuperNode began to search the entire game tree down to a certain fixed depth. That depth was configurable, and in practice, we searched until the number of leaf nodes uncovered exceeded a certain threshold. The threshold depended on the number of participating PeerNodes in the network.

At this stage, the leaf nodes ('work units') are delivered to the task management section of the SuperNode. Each node is given a priority, based on its importance derived from the results of the local search.

They are also assigned with an estimated value indicating the computational complexity, based initially on an estimate calculated from the statistical branching factor and the required search depth.

As the search progresses, this complexity is improved by calculating it as a function of the time it previously took to search that node, and the estimated branching factor within that search branch. If a node's estimated complexity is above a certain threshold value then it is further subdivided into its component child nodes.

The leaf nodes are distributed as work units to the PeerNode computers located throughout the world. Because of the potential for rogue PeerNode computers, the SuperNode was instructed to redundantly distribute information among participating systems. In an early version, we had a software flag, which controlled the level of duplication. The larger the number of nodes dedicated to each work unit, the more certain we could be that a result would be returned in a timely fashion, however it would involve an increasing waste of resources. A trade off was required, at least for this initial stage. This value conventionally was set to four, allowing four machines to process each individual work unit, but consequently reducing the number of resources available by the same factor.

In future work we would like to investigate the extension of this algorithm to a more intelligent algorithm that could adaptively distribute work units based on the expected importance of each node. Thus, important nodes might be distributed four times, but nodes which are judged as likely to be irrelevant may only be distributed once. That relevance could be calculated based on the heuristic move ordering algorithms, and the results of previous searches (if any). Though we want to avoid wastage, the absolute worst situation is when the SuperNode is left waiting on the result of one single node which has been distributed to an exceptionally slow (or unreliable) machine. The SuperNode is calibrated to test for inactivity after a certain time, and if there is still work to be done, then the pending jobs are distributed to the available idle machines, up to a certain maximum duplication factor.

As results are sent back to the SuperNode they are incorporated into the search. The core of the APHID algorithm involved passing periodically over the search tree root (the search tree subdivided down to the level of the leaf nodes) and incorporating new information as and when it arrives. This new information might lead to a cut-off at certain nodes, in which case a message is passed to the task management module of the SuperNode informing it that certain work units are no longer relevant. The SuperNode then contacts the relevant PeerNodes and sends them an 'abort' message.

The search ends when the list of pending units is exhausted, hence every root branch of the search tree has been assigned either an exact *minimax* value, or an insufficient upper bound. In this case, the tree can be fully evaluated, giving a 'best move' recommendation for the search position. If time allows then the search is extended one more ply, in which case the leaf node depth requirements are incremented by one and the new complexity is calculated. The leaf nodes are divided into their component leaf nodes if required. The APHID algorithm continues as above.

3 ChessBrain Architecture

The ChessBrain architecture involved a two-layer architecture based on a single central server communicating directly with a network of external contributors through TCP/IP connections.

The SuperNode is responsible for both coordinating the communications for the entire network, and coordinating the actual chess tree search algorithm. The individual contributors (PeerNodes) are treated as independent entities, connecting only to the SuperNode and not to each other. In addition to these components, the SuperNode also communicated with an external game server, on which the actual moves were played. This layout is illustrated in figure 2. In this figure, the game server represents the environment in which the game

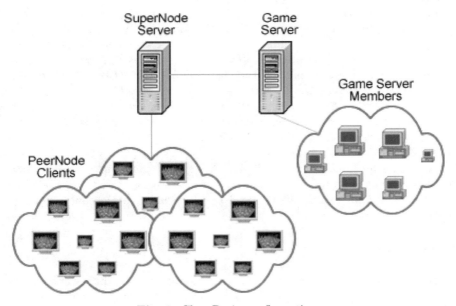

Fig. 2. ChessBrain configuration

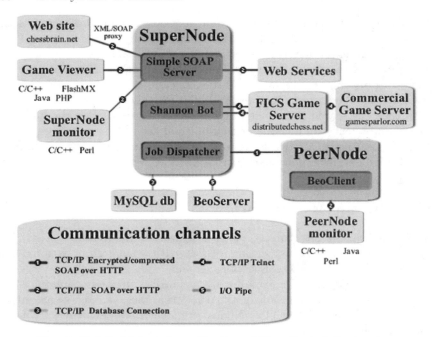

Fig. 3. Detailed internal specification of ChessBrain infrastructure

is actually played, and which communicates directly with the SuperNode to update the current game state whenever the opponent plays a move.

Each of these constituents of ChessBrain consisted of a number of complex components, illustrated in figure 3. In this diagram, you can see how the central server contained a number of individual components, including the game server interface mentioned previously, as well as the individual parts of the chess intelligence, namely the chess engine itself (BeoServer), the job dispatching algorithms and the database of pending work units.

Also connected to the central SuperNode server were a number of other services, including the entire website (which was updated in real time with the results of each move) and the monitoring software which we used internally to check the progress of each calculation.

Finally, the links to the individual PeerNode contributors, which were encrypted and compressed in order to increase security, reduce bandwidth requirements and maximize the speed at which information could be transferred.

3.1 Chess Interface

Leaf nodes are distributed to PeerNodes as work units. These encode the current position to be analyzed and the depth to which it should be searched. Work units are distributed to the connected PeerNodes on a request basis, though they are also ranked in order of estimated complexity using intelligent extrapolation from their recorded complexity at previous, shallower depths. In this way, the most complex work units can be distributed to the most powerful PeerNodes. Work units that are estimated to be far too complex to be searched within a reasonable time are further subdivided by one ply, and the resulting, shallower child nodes are distributed instead.

If a node in the parent tree returns a fail-high (beta-cut) value from a PeerNode search, then the remainder of the work units from that branch are pruned. This indicates that the position searched by the PeerNode proved very strong for the opponent, and therefore that the parent position should never have been allowed to arise. In this situation, we can cease analysis of the parent position and return an approximate upper limit for the score. PeerNodes working on these work units receive an abort signal, and they return immediately to retrieve a new, useful work unit.

4 Issues with Parallel Chess

We have already mentioned that parallelized chess suffers from a number of issues that do not affect the sequential version. The most important of these, and the first we'll consider, is the lack of shared memory. But there are many other considerations, some far less obvious than others, which we had to resolve to ensure the project's success.

4.1 Lack of Shared Memory

In parallel search, the PeerNodes cannot share each other's hash table data as we have no shared memory architecture and the volume of information is too large to be accurately transmitted within the time frame required. In ChessBrain, we developed a number of methods to compensate for this weakness, though it must be admitted that it is still in the large part insurmountable.

The most obvious advantage of shared memory is that it allows an efficient transposition table. With distributed computation, this is no longer the case, though each individual machine is able to create its own transposition table for each individual work unit. These tables are cleared at the end of each search process, in order to ensure that the results are reproducible and stable. Stale hash table entries are a major cause of extremely complex debugging issues.

The second drawback of this distributed infrastructure is that alpha-beta values, once updated, are not automatically available to all PeerNodes simultaneously. Now the fact that we will analyse some positions which later prove to be irrelevant is unavoidable. This is a consequence of the parallelism of the game tree search prohibiting the alpha-beta algorithm from working properly.

In summary, the alpha-beta algorithm stores values corresponding to the 'cut-off' scores in the current branch being searched. The cut-off score refers to the value below which a search result is uninteresting, and the value above which we can safely abandon the current line of enquiry. These values are honed and improved throughout the search, allowing increasingly substantial pruning to take place. However, once parallel search is employed then sibling branches are searched concurrently, before the initial alpha-beta values have been updated and hence the parallel searches cannot effectively be pruned.

The only possible partial solution to this problem is to update alpha-beta values as and when they are altered in any one branch. For every node that is sent out as a work unit, the initial alpha-beta values are transmitted along with the position and depth information. If those values are subsequently updated at the SuperNode, then they are transmitted to the PeerNode separately and the PeerNode continues with, or abandons its search where necessary.

4.2 Security

Chess tree search is an inherently fragile problem. If just one node is incorrectly evaluated, then it can plausibly disrupt the conclusion of the entire search process. When distributing work units to participants over the Internet, we must be absolutely sure not only that the participant receives the work unit exactly as planned, but also that the participant's computer evaluates that work unit accurately and without intervention, and that it transmits the true answer back to the central server reliably and accurately.

In order to overcome this challenge, we implemented secure encryption within the SuperNode and PeerNode software. Information is first compressed using Zlib compression and subsequently encrypted using the AES-Rijndael cipher.

In order to ensure the accuracy of the PeerNode game tree analysis, we removed all loadable data files from the PeerNode software including endgame databases, opening books and configurable parameter files. These were either abandoned entirely (in the case of the opening and endgame databases) or securely embedded into the executable file (in the case of the configuration data).

In order to avoid communication interception on the PeerNode machines, the chess playing engine and the communications software were integrated into a single software package, which communicated only with the SuperNode directly.

4.3 Recruitment

We had to recruit operators to join our cause and to dedicate their valuable CPU cycles to the ChessBrain project instead of any one of a number of other projects. Some of this could be achieved by word of mouth, but the most successful method was to encourage well-known Internet news sites to write up our project as a front-page article. We were included on the front page of Slashdot (http://slashdot.org) on two separate occasions, both of which resulted in our web servers straining under the load of a very substantial number of attempted connections. These two events enormously increased the public interest in our project, and resulted in contact from a number of distributed computing groups.

4.4 Community Relations

Collecting individual contributors is only one part of the problem facing the project manager for a distributed computation project. One of the most important sources of contribution is from so-called 'distributed computing teams', comprising of a number of individuals, each of which dedicates a certain amount of hardware in order to increase the amount of work completed by the team as a whole. Individual teams are not only very competitive internally, but they also compete against rival teams in an attempt to rise to the top of publicly visible leader-boards. Distributed computing projects benefit from this rivalry and thus encourage participants to dedicate as many resources as possible for as long as possible in order to generate the most substantial contributions for their team.

So, given the importance of distributed computing teams, it was of utmost importance to entice them to join the ChessBrain project, and to ensure that they contributed to the project as much as possible. One of the best ways of doing this was to give them unparalleled access to data and statistics about the performance of their team compared to other rivals.

Our main website was updated to show a short description of the major contributors, together with a small image of their choice and a link to their home page. In the statistics section, we included the number of nodes calculated by all the top contributors, and the top DC teams. The top team within the Copenhagen game contributed a total of 23,543,851,081 node evaluations to the overall effort. At an approximate rate, using contemporary hardware, of 100,000 nodes per second, this equates to over 64 hours of computing time, or 32 machines contributing 100% for the entire 2 hour match.

In practice, given the inefficient use of PeerNode resources, and the technical issues that we suffered in the match, the total potential contribution from this one team alone was probably many times greater than the theoretical value calculated above.

5 ChessBrain II

Though the initial ChessBrain project was completed in January 2004 with the world record attempt, it was decided to investigate the extension of the project to a second stage, ChessBrain II.

5.1 Motivation

The motivation behind ChessBrain II was quite simply to fix the most substantial problems with which we were presented in ChessBrain I:

- To enable far greater scalability in terms of number of contributors.
- To improve the overall efficiency by reducing excessive communication.
- To encourage recruitment of more participants.
- To improve the reliability of the chess tree search
- To improve the adaptability and 'intelligence' of the chess tree search

Whilst ChessBrain I was able to support over 2,000 remote machines, albeit with great difficulty, the lessons learned from the original design have enabled us to develop a vastly improved infrastructure, which goes a long way to removing the constraints of our first version.

5.2 Infrastructure

ChessBrain II, though superficially similar to its predecessor in some areas, actually rests upon a completely novel, custom-built server application, called msgCourier. This enables the construction of a hierarchical network topology that is designed to reduce network latency through the use of clustering as outlined in figure 4. The resulting topology introduces network hubs, the importance of which to graph theory has also been well covered in research superseding the random graph research of Erdos and Renyi and in the social network research of Milgram. In brief, well placed communications hubs help create small world effects which radically improve the effectiveness of networked communication. [8, 9].

The ChessBrain II system consists of three server applications, a SuperNode, ClusterNode and PeerNode. The PeerNodes retain their original functions, accepting work units, searching the resultant game trees to the required depth. PeerNodes now communicate with a designated ClusterNode, rather directly with a SuperNode. The role of the SuperNode has changed significantly in that is only communicates with ClusterNodes which in turn communicate with PeerNodes.

The central server no longer distributes work units directly to the PeerNodes, as was the case with ChessBrain I, instead work units are sent to an array of first-level ClusterNodes, operated by trusted community members. These ClusterNodes contain no chess-playing code and behave as network hubs (relay points) through which the complete set of work units can be passed.

Each ClusterNode contains a complete listing of all PeerNodes connected to it, together with a profiling score to determine the approximate CPU speed of each PeerNode, exactly as was implemented in ChessBrain I. Each PeerNode connects to one and only one ClusterNode, and direct communication with the SuperNode is forbidden.

The ChessBrain Project – Massively Distributed Chess Tree Search

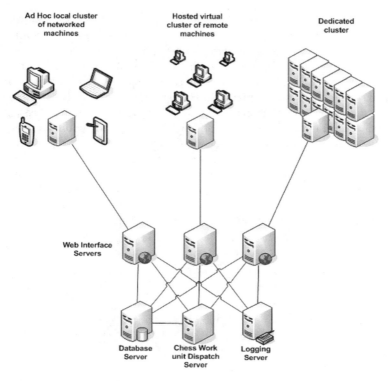

Fig. 4. ChessBrain II configuration

Table 1. Server Types

Component	Purpose
SuperNode	Central server. Interfaces with the actual game being played. Manages work unit creation, partitioning and distribution. Handles generation of global statistics.
ClusterNode	Manages communities of local and distributed PeerNode servers. Receives instructions from the SuperNode. Handles generation of local cluster statistics.
PeerNode	Compute node server. Performs work unit processing. Reports directly to a chosen ClusterNode.

The ClusterNodes, having been allocated a selection of individual work units by the SuperNode, then divide up these work units as they see fit based on the profiling data that they obtain from their own network of PeerNodes. The primary considerations are that the work units should be distributed to sufficient machines to ensure a reliable reply within the time required, plus to ensure that the work units perceived to require a greater computation effort are allocated to those PeerNodes deemed most fit to analyse them. The algorithm used here is the same as that used by the SuperNode in order to decide which work units should be sent to which ClusterNodes.

Future plans involve allowing the ClusterNodes not only to act as relay hubs, but actually to take part in the chess calculation, by passing them only a handful of root positions, and allowing them to search, partition and distributes the component leaf nodes within those positions themselves. This further reduces the communications overhead between the SuperNode and the ClusterNodes, as only one position need be sent instead of the component leaf nodes from that position's game tree.

Once the calculation is completed, the ClusterNodes will then return a single result to the central SuperNode, instead of many.

5.3 ChessBrain II Communication Protocols

Early versions of ChessBrain relied on industry standard XML data encoding first using XMLRPC, and later using SOAP. The decision to use SOAP was driven by a desire for interoperability with emerging web services. However, the need to streamline communication has steered us toward minimizing our use of XML in favour of economical string based S-Expressions [10]. When sending thousands of work units to a significant number of remote machines, the size of an individual message is paramount.

To further streamline communication we've implemented a compact communication protocol similar to the Session Initiation Protocol (SIP) [11] for use in LAN and cluster environments where we favour the use of connectionless UDP rather than stream-based TCP communication.

The ChessBrain I communication protocol consisted of XML content which was first compressed using Zlib compression and then encrypted using the AES Rijndael cipher. Although each PeerNode was quickly able to decrypt and decompress the payload content, the SuperNode server had a significantly more demanding workload, having to encrypt and decrypt not one, but thousands of messages for each move.

With ChessBrain II we've eliminated direct PeerNode communication with the central SuperNode and introduced the concept of batch jobs, which combine multiple jobs into a single communication package. The reduction in the number of messages that need to be sent reduces the communication load, as well as the impact to the TCP stack. The grouping of jobs also greatly improves the compression ratio of transmitted packages, meaning that the overall bandwidth required at the SuperNode is significantly reduced.

5.4 Architecture Advantages

The most significant architectural change to ChessBrain involves the introduction of network hubs called ClusterNodes, as outlined earlier.

ChessBrain I used a single SuperNode server to coordinate all communication within a network of thousands of machines. Each dispatched job required a direct session involving the exchange of multiple messages between the SuperNode and its PeerNode clients. With ChessBrain II, jobs are distributed from a central server at *distributedchess.net* to remote ClusterNodes, which in turn manage their own local communities of PeerNodes. Each ClusterNode receives a batch of jobs, which it can directly dispatch to local PeerNodes thereby eliminating the need for individual PeerNodes to communicate directly with the central server. As we increasingly utilize the power of substantial compute clusters, this method becomes vital. Each ClusterNode collects completed jobs and batches them for return shipment to the central SuperNode server. The efficient use of ClusterNode hubs results in a reduced load on the central server, efficient use of compute clusters, reduced network lag, and improved fault tolerance.

The expected ClusterNode operators will largely be individuals desiring to connect local machines as an *ad hoc* cluster. Indeed during the use of ChessBrain I we detected locally networked clusters containing between five and eighty machines. Most local networks in existence today support connection speeds between 10 to 1000 MBit per second, with the lower end of the spectrum devoted to older wireless networks, and the higher end devoted to corporate networks, research networks and compute clusters. ChessBrain II is designed to utilize cluster machines by taking full advantage of local intranet network speeds and only using slower Internet connections to communicate with the SuperNode when necessary.

If we assume that there are roughly as many PeerNodes connected to each ClusterNode as there are ClusterNodes, then the communications costs for each Cluster node, and indeed the SuperNode itself, is reduced to its square root. So, with total node count N, instead of one single bottleneck of size N, we now have approximately (sqrt(N)+1) bottlenecks, each of size sqrt(N). When addressing scalability issues, this is a definite advantage, allowing us to move from an effective node limit of approximately 2,000 up to around one million machines.

5.5 Architecture Drawbacks

It is only fair to consider the drawbacks of the above architecture and to explain why it may not be suitable for every plausible application.

Firstly, as with any distributed computation environment, there is a substantial overhead introduced by the remote communication itself. Communication costs increase as the number of available remote machines increases. For each PeerNode, the SuperNode (in ChessBrain I) or the ClusterNodes (in ChessBrain II) will need to generate work unit packages and distribute them, encrypted and compressed, over the network. ChessBrain I involved a single server solution that was overburdened as an unexpectedly large number of remote machines became available. Communication overhead on ChessBrain

I reached approximately one minute per move under peak conditions. However, with the experience gained since that first exhibition match, and with the subsequent redesign of ChessBrain I, we have reduced the overhead to less than ten seconds per move even with preliminary versions of ChessBrain II. This is a very substantial saving, and we anticipate that it will be still further reduced as development continues.

The presence of a measurable communication overhead means that shorter time scale games are not currently suitable for distributed computation. 'Blitz' games are conventionally played with a total time limit per player of either five or ten minutes, which gives approximately ten seconds per move once the opening has been played. Even with the current system, that would be entirely eaten up with communications overhead, not giving any processing time whatsoever.

However, games that favour a higher quality of play over speed of play are likely to make good use of distributed computation. Correspondence games, for example, or those for which the main purpose of distributed computation is analyzing the standard theoretical results. For example, with Chess it would be interesting to search the opening board position with a substantial distributed computation effort in order to see if any improvement can be made on conventional opening book moves.

Anyone who has ever attempted to write a fully-functioning alpha-beta pruning chess search algorithm featuring a multitude of unsafe pruning algorithms such as *null-move pruning*, will immediately appreciate the complexity of debugging a search anomaly produced from a network of several thousand computers, each of which is running a number of local tree searches and returning their results asynchronously. Some of the complexities of such an approach are covered in earlier work [12]. The accumulation of debugging information from remote machines, some of which may become unavailable unexpectedly during the search, is a very substantial issue. Likewise, for the transmission and storage of enormous debugging log files over the Internet, not to mention their eventual analysis. Most initial development with ChessBrain I and II was carried out internally with a local compute cluster to which we had direct access.

Adding hierarchical distribution increases complexity, and highlights the importance of considering how a distributed application will be tested early in the design phase. With ChessBrain II we've had to build specialized testing applications in order to identify and correct serious flaws, which might otherwise have escaped detection. Such a suite of testing tools is invaluable for a distributed application of this size. We also recommend the use of iterated testing – that is, starting with a very simple test situation, and verifying that the correct output is produced, then generating increasingly more complex tests, at every stage verifying that the correct output is produced. Whenever any code change is committed, regression tests should be executed to identify deviations from expected results.

5.6 Comparison with Alternative Parallel Algorithms

Other approaches towards parallelising search problems focus primarily on individual, co-located compute clusters with shared memory and identical (homogeneous) architecture throughout. The ChessBrain project has many advantages and disadvantages over many other distributed computation projects. Locally parallel search on a large supercomputer is clearly vastly superior, should one have access to a suitable location in which to host the cluster and, more importantly, the substantial funding required to purchase and to operate it in the first place. We acknowledge that such levels of funding are rarely available to game enthusiasts, or even to academic researchers.

The main advantages of the fully distributed volunteer-computing method over that used by, for example, the Deep Blue project [5] and the more recent Hydra project [6] are as follows:

- **Processing power** – With many entirely separable applications, parallelising the search is a simple way to get extra processing power for very little extra overhead. For chess, the parallelisation procedure is highly inefficient when compared to serial search, but we chose this application because of its inherent difficulty, our own interest and its public image.
- **Distributed memory** – With many machines contributing to the search, the total memory of the system is increased massively. Though there is much repetition and redundancy, this still partly overcomes the extra practical barrier imposed by the finite size of a transposition table in conventional search.
- **Availability** – the framework described in this paper is applicable to a wide range of projects requiring substantial computing power. Not everyone has access to a supercomputer or a substantial Beowulf cluster.
- **Costs** – It's easier to appeal to 10,000 people to freely contribute resources than it is to convince one person to fund a 10,000 node cluster.

Drawbacks include:

- **Communication overheads** – time is lost in sending/receiving the results from PeerNodes.
- **Loss of shared memory** – In games such as chess, the use of shared memory for a transposition table is highly beneficial. Losing this (amongst other cases) introduces many overheads into the search time [13]
- **Lack of control** – the project manager has only a very limited control over whether or not the contributors choose to participate on any one occasion.
- **Debugging** – This becomes horrendously complicated, as explained above.
- **Software support** – The project managers must offer support on installing and configuring the software on remote machines.
- **Vulnerability** – The distributed network is vulnerable to attacks from hackers, and must also ensure that malicious PeerNode operators are unable to sabotage the search results.

At the present time, we are not aware of any other effort to evaluate game trees in a distributed style over the Internet.

5.7 Comparison with Other Chess Projects

We are often asked to compare ChessBrain with famous Chess projects such as Deep Blue and the more recent standalone, commercially available single processor software. A direct comparison is particularly difficult as ChessBrain relies on considerably slower communication and commodity hardware. In contrast, Deep Blue was based on a hardware-assisted brute force approach. A more reasonable comparison would be between distributed chess applications running on GRIDs and distributed clusters.

However, we want to move away from comparisons because our vision of the ChessBrain project was not purely to produce a strong chess-playing engine, but rather to create a demonstration platform whereby we could prove the feasibility of high-speed, distributed computation with an inhomogeneous network of contributors connected over the Internet. The techniques in this work are just as applicable to many other search problems within the field of gaming, and outside. However, we envisage eventually connecting ChessBrain II to an international chess server where competitors will be able to challenge it freely in tournament time control games. We also plan to do some fundamental analysis of problematic test positions and opening book theory, where our time control can be set to any value required.

5.8 msgCourier

While considering architectural requirements for ChessBrain II, we investigated a number of potential frameworks including the Berkeley Open Infrastructure for Network Computing (BOINC) project. BOINC is a software application platform designed to simplify the construction of public computing projects and is presently in use by the SETI@home project, CERN's Large Hadron Collider project and other high-profile distributed computing projects [14].

After testing the BOINC infrastructure for some time, we decided that the only way to satisfy ChessBrain's unique requirements would be to construct a new underlying server application technology from scratch [15, 16, 17]. One of our requirements for ChessBrain II's software is that it must be a completely self-contained application that is free of external application dependencies. In addition, our solution must be available for use on both Microsoft Windows and Unix-based platforms, while requiring near zero configuration. The rationale behind these requirements is that ChessBrain II allows some of our contributors to host ClusterNode servers and it is critically important that our contributors feel comfortable with installing and operating the project software. Also, given the limited resources of the ChessBrain project, we couldn't commit to offering any but the most basic of support facilities. We found that

BOINC requires a greater level of system knowledge than we were realistically able to require from our contributors without losing a significant fraction of them.

Lastly, BOINC was designed with a client and server methodology in mind, while our emerging requirements for ChessBrain II include Peer-to-Peer functionality. Though the current configuration does not use this method, we have plans in place for introducing it at a later point in time.

Well over a year ago we began work on the Message Courier (msgCourier) application server in support of ChessBrain II. MsgCourier is designed to support speed critical computation using efficient network communication and enables clustering, which significantly improves overall efficiency. Unlike other technologies, msgCourier is designed to enable the use of local machine clusters and to harness as efficiently as possible existing Beowulf clusters.

MsgCourier is a hybrid server application that combines message queuing, HTTP server and P2P features. When we embarked on this approach there were few such commercial server applications. Since that decision was made, Microsoft has released SQL Server 2005, which combines a SQL Engine, HTTP server and messaging server features. The industry demands for performance necessitates the consideration of hybrid servers. The msgCourier project is distinct from this commercially available offering because it is both opensource in nature, and freely distributed.

We chose to build msgCourier independently of ChessBrain (and free of chess related functionality) in the hopes that it would prove useful to other researchers keen to embark on similar projects for themselves. In fact, the msgCourier project is so general that its applications spread far beyond those of distributed computation.

The following were a few of our primary design considerations:

- A hybrid application server, combining message queuing and dispatching with support for store and forward functionality.
- Multithreaded concurrent connection server design able to support thousands of simultaneous connections.
- High-speed message based communication using TCP and UDP transport.
- Built-in P2P functionality for self-organization and clustering, service adverting and subscribing.
- Ease of deployment with minimal configuration requirements.
- Built-in security features which are comparable to the use of SSL and or SSH.

The msgCourier project is under continued development. We are keen to emphasize here that the relevance of the ChessBrain project is not just to the specific field of computer chess, but also to any distributed computation project. Hence, we believe that the msgCourier software is a valuable contribution to all areas of computationally intensive research. The direct application here demonstrates that the framework is also flexible enough to operate within gaming scenarios, where results are required on demand at high speed and with high fidelity, often in highly unpredictable search situations.

More information on the msgCourier project is available at the project home page:

http://www.msgcourier.com

6 Conclusions

The ChessBrain project investigated the feasibility of massively parallel game tree search using a diverse network of volunteers connected via the Internet. Game tree search is an especially difficult problem because games almost always have a fixed time limit imposed upon them, which means that a 'good enough' answer must be returned before the clock runs down, regardless of whatever problems might afflict network connectivity, participant numbers and tree complexity.

The challenges affecting the ChessBrain project were many, though they generally fall under the following headings:

- **Sequential algorithms** – How to adapt algorithms such as alpha-beta, which were designed to optimise sequential search, in order to ensure that they work as well as possible with parallel search.
- **Communication bandwidth** – How to maximise the speed with which work units are distributed to the contributing PeerNodes, and the efficiency with which the resultant analysis is collected back at the SuperNode.
- **Work distribution** – How to distribute the work efficiently to the available resources without wasting excess effort on irrelevant nodes, or relying too heavily on the efficiency of individual contributors.
- **Debugging** – How to understand the workings of a distributed system with debugging files stored with the PeerNodes, not the central server.
- **Security issues** – How to ensure that the search is not invalidated by corrupted results coming from compromised (malicious) contributors, from man-in-the-middle attacks, or from transmission errors.
- **Social issues** – How to convince a sufficient number of contributors to join up.

Whereas ChessBrain I addressed these challenges to a degree, especially the last two, its successor, ChessBrain II is aiming to significantly reduce the inefficiencies experienced by its predecessor.

The technologies displayed in the ChessBrain project have more serious applications in the field of distributed computation, wherever a project is unable to run on a single, homogeneous cluster of nodes. Especially, this technology applies to areas such as financial and military applications, where a large-scale distributed calculation needs to be made within a fixed or variable time limit.

For example, within finance, the evaluation of exotic financial derivatives is often accomplished using a statistical Monte Carlo technique, which can easily be parallelised. However, in this case, time is also important because evaluation of a derivative product often needs to be accomplished within a strict timeframe. Often, investment banks have large, dedicated computing clusters in order to perform such calculations, usually with thousands of nodes. However, such clusters are often run inefficiently, with poor job distribution and an inefficient management of the variability inherent in the system, both in hardware and in transient network issues.

The parallels with the field of military planning are obvious, especially from a game such as chess. The main difference in this case is that real-world military strategy is not turn-based. However, though there is no fixed deadline for strategies to be deployed, time is almost always of the essence. The problem with expanding our research to such a critical application, however, is mainly one of security: in both these examples (banking and military planning), the information being processed is of very substantial importance and intrinsic value. Many of the skills we learned through the ChessBrain project concerned the management of a distributed team of unknown contributors, with unknown hardware and software configurations and, at least potentially, a desire to disrupt the experiment. In this situation, the ability to understand and mitigate human problems was almost as important as the ability to design and implement efficient algorithms.

For more information, or if you would like to volunteer to help in the ChessBrain project, please visit the project websites at the following URLs:

http://www.chessbrain.net
http://www.distributedchess.net

References

1. Shannon, C.E.: "Programming a Computer for Playing Chess", Philosophical Magazine, Ser.7, Vol. 41, No. 314, March (1950)
2. Beal, D. F. *"Advances in Computer Chess 2,"* Chapter "An Analysis of Minimax", pp. 103–109. Edinburgh University Press, (1980)
3. Fraenkel, A. S. & Lichtenstein, D., "Computing a perfect strategy for n*n chess requires time exponential" *Proc. 8th Int. Coll. Automata, Languages, and Programming*, Springer LNCS 115 (1981) 278–293 and *J. Comb. Th. A* 31 199–214. (1981)
4. Newell, A., Shaw J.C., & Simon, H.A. "Chess-playing programs and the problem of complexity." *IBM J. Res. Develop.* 2:320–25, (1958)
5. Campbell, M., Joseph Hoane Jr., A., Hsu, F., "Deep Blue", Artificial Intelligence 134 (1–2), (2002)
6. http://www.hydrachess.com
7. Brockington, M., "Asynchronous Parallel Game Tree Search", PhD Thesis, University of Alberts, (1997)

8. Barabasi, A-L. "Linked: The new Science of Networks", Cambridge, MA: Perseus (2002)
9. Gladwell, M. "The Tipping Point", Boston: Little and Company (2000)
10. Rivest, R. L., "S-Expressions"., MIT Theory group, http://theory.lcs.mit.edu/~rivest/sexp.txt
11. Session Initiation Protocol (SIP) http://www.cs.columbia.edu/sip/
12. Feldmann, R., Mysliwietz, P., Monien, B., "A Fully Distributed Chess Program", Advances in Computer Chess 6, (1991)
13. Feldmann, R., Mysliwietz, P., Monien, B., "Studying overheads in massively parallel MIN/MAX-tree evaluation", Proc. 6th annual ACM symposium on Parallel algorithms and architectures, (1994)
14. Berkeley Open Infrastructure for Network Computing (BOINC) project: http://boinc.berkeley.edu/
15. Justiniano, C., "Tapping the Matrix", O'Reilly Network Technical Articles, 16th, 23rd April 2004. http://www.oreillynet.com/pub/au/1812
16. Justiniano, C., "Tapping the Matrix: Revisited", BoF LinuxForum, Copenhagen (2005)
17. Lew, K., Justiniano, C., Frayn, C.M., "Early experiences with clusters and compute farms in ChessBrain II". BoF LinuxForum, Copenhagen (2005)

Resources

Books

- Beal, D.F. (Ed.), "Advances in Computer Chess 5", 1989, Elsevier Science Publishing, ISBN 0444871594.
- Frey, P.W., "Chess Skill in Man and Machine", 1983, Springer, ISBN 0387907904.
- Heinz, E. A., "Scalable Search in Computer Chess: Algorithmic Enhancements and Experiments at High Search Depths", 2000, Morgan Kaufmann, ISBN 3528057327.
- Hsu, F-H, "Behind Deep Blue: Building the Computer that Defeated the World Chess Champion", 2002, Princeton University Press, ISBN 0691090653.
- Levy, D. & Newborn, M., "How Computers Play Chess", 1991, Computer Science Press, ISBN 0716781212.

Organizations

- International Computer Games Association (http://www.cs.unimaas.nl/icga/)

Originally the International Computer Chess Association (ICCA). This organisation runs regular, high profile international events in chess and other similar games, and publishes a quarterly journal.

- International Chess Federation (http://www.fide.com)

The governing body for (human) chess worldwide. It runs the official world championships.

Discussion Groups / Forums

- TalkChess Computer Chess Club (http://www.talkchess.com/forum/)

    Frequented by many of the world's top experts in the field of computer chess.

- Computational Intelligence and Games (http://groups.google.com/group/cigames)

    Covering the topic of computational intelligence in a wide variety of games.

Key International Conferences

- Annual IEEE Symposium on Computational Intelligence and Games

(http://www.cigames.org/)
    Includes many of the world's top researchers in the field of computational intelligence in games.

Open Source Software

- Beowulf Chess Software (http://www.frayn.net/beowulf/)

    Colin Frayn's free, open-source chess engine, written in C and C++.

- WinBoard and Xboard (http://www.tim-mann.org/xboard.html)

    Tim Mann's free chess viewer software.

- WinBoard/Xboard Engine List (http://www.tim-mann.org/engines.html)

    A list of chess engines that work with the WinBoard/Xboard interface. Many of these are free & open source.

Databases

Endgame tablebases (ftp://ftp.cis.uab.edu/pub/hyatt/TB/)
    Download the 3,4,5-piece endgame tablebases, which show correct play for any endgame position with very few pieces left on the board. Currently, some 6-piece tablebases exist. The piece count includes the two kings.

Designing and Developing Electronic Market Games

Maria Fasli and Michael Michalakopoulos

University of Essex, Department of Computer Science, Wivenhoe Park, Colchester CO4 3SQ, United Kingdom
{mfasli,mmichag}@essex.ac.uk

Abstract. As software agents are well-suited for complex, dynamic and constrained environments that electronic markets are, research into automated trading and negotiation has been flourishing. However, developing electronic marketplaces and trading strategies without careful design and experimentation can be costly and carries high risks. This chapter discusses the need for tools to support the design and implementation of electronic market simulations or games to emulate real life complex situations of strategic interdependence among multiple agents. Such games can be used to conduct research on market infrastructure, negotiation protocols and strategic behaviour. After a brief introduction into the field of agents, we present an overview of agent negotiation in general, and auction protocols in particular, which are among the most popular negotiation protocols. We then present the e-Game platform which has been developed to support the design, implementation and execution of market simulation games involving auctions. How the development of market games is facilitated is demonstrated with an example game.

1 Introduction

The vision of future electronic marketplaces is that of being populated by intelligent software entities – agents – representing their users or owners and conducting business on their behalf. Unlike "traditional" software, agents are personalized, semi-autonomous and continuously running entities [20] and these characteristics make them ideal for complex, and unpredictable environments that electronic markets are. The deployment of software agents in electronic commerce would bring about significant advantages: it would eliminate the need for continuous user involvement, reduce the negotiation time and transaction costs and potentially provide for more efficient allocation of goods and resources for all parties involved. As a result, research into agent-based and multi-agent systems for automated trading has intensified over the last few years.

Exchanges (electronic or otherwise) involve broadly speaking three main phases: firstly potential buyers and sellers must find each other or meet in a marketplace, secondly they need to negotiate the terms of the transaction and finally they execute the transaction and the goods/monetary resources change hands. Agent technology can be used in all three phases, from identifying potential trading partners, to negotiation and payment systems. Agents can negotiate for goods and services on behalf of their users (individuals or organizations) reflecting their preferences and perhaps even negotiation strategies.

However, despite their increased popularity, the huge potential of agents and multi-agent systems has not been fully realized. The reasons for the slow uptake of agent technology in particular with regards to e-commerce applications are manifold. Firstly, there is a lack of standardization that permeates the field of agents and multi-agent systems. This inevitably raises concerns as individuals and organizations would like to use standard and stable technologies. Secondly, there are inherent issues with trust. Trust becomes very important if an agent's actions can cause its owner financial, or even psychological harm. The prospect of disclosing personal, financial or otherwise sensitive information to an agent and delegating to it the task of conducting business on one's behalf does involve a number of risks. Thirdly, although implementing simple agents to carry out single negotiation tasks is easy, developing flexible agents that can operate in highly dynamic and uncertain environments is nontrivial. In practice, agents may have to negotiate for a bundle of perhaps interrelated goods being traded via different negotiation protocols. Developing agents that can use different negotiation protocols and compete effectively in complex markets with complementary and substitutable goods is a nontrivial challenge. The successful performance of an agent does not only depend on its strategy, but critically on the strategy of the other agents as well. Designing efficient and effective decision-making algorithms and strategies is difficult since real world data about human traders are hard to obtain.

This chapter discusses one way that the last problem can be addressed. In particular, we discuss the need for tools to support the design and implementation of electronic market simulations or games for conducting research on market infrastructure, negotiation protocols as well as strategic behaviour. We present a platform that provides the facilities for developing electronic market games that involve a variety of auction protocols. How the design and implementation of market games is aided is described and demonstrated with an example game. The rest of the chapter is organized as follows. First we provide a brief introduction into agents and multi-agent systems to put the rest of the sections into context. Next we provide an overview of agent negotiations and an introduction into auctions. The following section discusses the need for developing market games. A discussion of the relevant work in the literature with regards to electronic marketplaces and trading agent platforms follows. The architecture and the most significant features of the e-Game auction platform are described next. The following section discusses game development and the facilities provided by e-Game to facilitate this.

The development of a simple game is presented next. The paper ends with a discussion on further work and the conclusions.

2 Agents and Multi-agent Systems

Agents and Multi-agent systems are a relatively new area of research and development in Computer Science. From a historical point of view, the area has its roots in Artificial Intelligence (AI) and Distributed Artificial Intelligence (DAI). It developed as a separate strand of research in the 1990's, as it was recognized that agents and multi-agent systems can play an important role in the development of distributed open systems. Nowadays, the discipline is very much interdisciplinary and encompasses concepts and ideas from other disciplines as diverse as mathematics, sociology, philosophy, computer science and economics – though this is not an exhaustive list.

Although there is no universal agreement as to what exactly constitutes an agent, most researchers and practitioners understand agents to be computational systems which may consist of both hardware and software components which exhibit characteristics such as autonomy, proactiveness, reactiveness and sociality [10]. Autonomy enables an agent to have control over its execution and behaviour, and act without the direct intervention of the user or other entities. Proactiveness enables an agent to actively seek to find ways to satisfy its design objectives or its user's goals. Reactiveness enables an agent to react to the changes that occur in its environment in a timely fashion and which affect its goals and design objectives. Finally, it is very rare that an agent will be useful on its own, therefore sociality enables an agent to do things with other agents by coordinating with them. Coordination can take two forms: cooperation and negotiation. In cooperation, agents usually have to work with each other as they cannot do something on their own, i.e. the task may be too complex for one agent. Agents are usually cooperative, and although they are interested in maximizing their own benefit, they are also interested in the welfare of the group as a whole. In negotiation, self-interested agents compete with each other for resources, goods etc. They are strictly utility maximizers, and they are not interested in the welfare of the other agents or the society.

Multi-agent systems comprise multiple agents who are situated in a common environment and they can interact with each other through communicative or physical actions. Interaction is both necessary and unavoidable in a common environment and agents may interact with one another because they have to or unintentionally. Interaction is shaped through a set of rules, namely an interaction protocol which dictates the messages or actions that are possible in an interaction situation. We can classify interaction protocols into three broad categories. Communication protocols enable agents to communicate with each other. Such protocols dictate the kind of messages that can be exchanged between agents and their meaning. Cooperation protocols enable agents to work together to achieve a common objective, perform a complex task or solve a difficult problem. Negotiation protocols enable agents to reach agreements when they are competing with each other and have conflicting goals.

Agents and multi-agent systems have found applications in a number of domains ranging from manufacturing control, information management, entertainment to education. More recently, research efforts have been focused on deploying agents in electronic commerce. Such agents represent individuals or organizations and they search for products or services, evaluate different alternatives, negotiate deals with other trading partners and execute transactions. Indeed, the potential for agents in e-commerce is enormous, and some researchers and technology analysts consider e-commerce to be the killer application for agents.

3 Negotiation Protocols

When agents are in competition or have conflicting goals and attempt to resolve these, they do not interact randomly, but their interaction is governed by a set of rules, namely a protocol [10]. A negotiation or market protocol provides a set of rules and behaviours to be followed by agents that will interact in it. *Mechanism design* is the design of protocols governing strategic interaction among self-interested agents. A negotiation situation is characterized by three elements:

1. The negotiation set which represents the space of possible offers or proposals that agents can make.
2. A protocol which defines the rules that the agents need to follow in order to arrive at an outcome and also rules on what constitutes a legal offer or proposal.
3. A collection of strategies that agents can use to participate in the negotiation. The agents' strategies are private, they are not dictated by the protocol itself and may take into consideration the possible strategies of other agents.

As negotiation situations may differ, some protocols may be more suitable than others for particular situations. There are a number of factors that determine the type of mechanism or negotiation protocol that can be designed for or employed in a particular situation:

– *Number of attributes.* Negotiation can take place over one attribute (e.g. price) or many (e.g. price, quantity and delivery day). Protocols that enable multi-attribute negotiation are more complex than those for single-attribute negotiation due to mainly two issues. Firstly, the negotiating agents have to be able to express preferences and construct offers on multiple attributes which complicates the individual agent's reasoning

mechanism. Secondly, determining the winner in such protocols is a difficult problem especially if multiple winners are to be allowed.
- *Number of agents*. The number of agents involved in the negotiation process also affects mechanism design. Negotiation can be on *one-to-one*, *one-to-many*, or *many-to-many*. In the last case, if there are n agents, there are potentially $n(n-1)/2$ concurrent negotiation threads active.
- *Number of units*. The price of multiple units may differ from protocol to protocol.
- *Interrelated goods*. In certain cases, the goods are only worth to agents when they are in combinations, thus agents negotiate over packages or bundles rather than individual goods. Protocols that enable negotiation over interrelated goods are more complex than those for single good negotiations as expressing and evaluating offers on combinations of multiple objects (which may be substitutable or interrelated) are difficult problems.

A number of protocols have been developed to deal with different negotiation situations. Bargaining protocols, for instance, enable agents to reach a deal through a process of making offers and counteroffers until a deal can be agreed on which is acceptable to all negotiating parties. Voting protocols take as input the individual preferences of a society of agents and attempt to aggregate them in a social preference which dictates an outcome to be imposed on the society. Auctions are a family of negotiation protocols whose distinctive characteristic is that the price of the good under negotiation is determined by the marker participants. We will examine auctions in more detail in subsequent sections.

4 Desiderata for Negotiation Protocols

From the multi-agent systems point of view mechanism design is the design of protocols governing multi-agent strategic interactions. Agents have their individual preferences and evaluations of goods and services that are available in the market and they are seeking to maximize their utilities by exchanging goods and services with the other participants. The aim of traditional mechanism design is to design a system in which rational agents interact in such a way so that desirable social outcomes follow. The desired properties of the social outcomes are encapsulated in a social choice function which ultimately describes what is to be achieved through the use of the mechanism [21], i.e. utility maximization across all agents, efficient allocation of goods and resources, etc.

When designing mechanisms for strategic interactions we are particularly interested in those mechanisms that enjoy certain game-theoretic and computational properties [29]:

Pareto efficiency. Pareto efficiency or economic efficiency measures global good. A solution x is Pareto efficient if there is no other solution x' such that at least one agent is better off in x' than in x and no agent is worse of in x' than in x. Although Pareto efficiency is a useful criterion for comparing the outcomes of different protocols, it has nothing to say about the distribution of welfare across agents.

Social welfare. Social welfare is measured through a welfare function which provides a way of adding together the different agents' utilities. A welfare function provides a way to rank different distributions of utility among agents, and thus provides a way to measure the distribution of wealth across agents. But to be able to measure social welfare, inter-agent utility comparisons are required which are not always straightforward and may involve utility transformations.

Individual rationality. Individual rationality dictates that an agent should not lose out by participating in a mechanism. Put differently, an agent should only participate in a mechanism if its payoff from the negotiated solution is no less than the payoff that the agent would have by not participating. A protocol is individual rational if it is individual rational for each participating agent.

Stability. Protocols can be designed with dominant strategies: strategies that an agent can use which guarantee maximization of payoffs. In particular, we prefer protocols in which truth-telling is the agents' dominant strategy as agents can decide what to do without having to counterspeculate about the others' strategies. A negotiation protocol is stable if it is designed so that it motivates the agents to behave in a desired way.

Budget balance: In strict budget balance the total payment that agents make must be equal to zero, in other words money is not injected into or removed from a mechanism. Alternatively, in weak budget balance the total payment is nonnegative, this ensures that the mechanism does not run at a loss. In *ex ante* budget balance the mechanism is balanced on average, while in *ex post* the mechanism is balanced at all times and for all instances.

Computational efficiency. The negotiation protocol should be computationally efficient. As little computation as possible should be required on behalf of the agents.

Distribution and communication efficiency. Distributed protocols are preferred to protocols with a central control point, since they avoid failure and performance bottleneck. Protocols should be communication efficient by requiring as little communication as possible to reach the desired outcome.

As we are interested in negotiation protocols for electronic markets computation is crucial. Computation in mechanism design can be distinguished along two distinct dimensions [27]:

Agent:

– Valuation complexity. How much computation is required to provide preference information within a mechanism?

- Strategic complexity. Are there dominant strategies? Do agents have to counterspeculate or do any opponent modelling when computing an optimal strategy?

Infrastructure/mechanism:

 - Winner-determination complexity. How much computation is required of the mechanism infrastructure in order to compute an outcome given the agents' preferences?
 - Communication complexity. How much communication is required between the agents and the mechanism to compute an outcome?

The challenge is to design computationally efficient mechanisms without sacrificing useful game-theoretic properties, such as efficiency and stability. Mechanism design has a fundamental role to play in devising protocols for complex distributed systems that comprise of self-interested interacting agents. In addition, the use of simulated marketplaces where the consequences of using different protocols on agent behaviour as well as strategic interactions can be studied in detail, can play an important role in designing efficient electronic markets.

5 Auctions

One of the most popular ways of negotiating for goods and services is via auctions [3, 17]. Auctions are a method of allocating goods based upon competition among the interested parties [10]. They constitute one of the oldest forms of market and some pinpoint their origin to Babylon in 500 BC. The term auction comes from the Latin root "auctio" which means "increase". Nowadays all sorts of goods and services are being traded in auctions ranging from paintings to spectrum licenses. With the advent of the World Wide Web, auctions have become extremely popular as the negotiation protocol of choice for conducting consumer-to-consumer (C2C) negotiations. Auction sites such as eBay [9] and onSale [26] have been reporting millions of dollars in transactions from auction sales. Governments have used auctions to sell spectrum and TV licenses, rights to drill for oil and for privatizing government owned companies.

There are two main self-interested parties in an auction: the auctioneer and the bidders. The auctioneer is usually a seller who wants to sell goods at the highest possible price (or subcontract out tasks at the lowest possible price) who may be the owner of the good or service, or a representative of the actual seller. An auctioneer can also be a buyer who is looking to buy a good from a number of sellers at the lowest possible price. The bidders can be buyers who want to buy goods at the lowest possible price (or get awarded contracts at the highest possible price), or sellers who compete for sales. Agents that participate in auctions are self-interested and rational seeking to maximize

Table 1. Auction terminology

Term	Meaning
Bid	Bids are offered by bidders to buy or sell the auctioned good.
Reservation price	The maximum (minimum) price a buyer (seller) is willing to pay (accept) for a good. This is usually private information.
Process bid	The auctioneer checks the validity of a submitted bid according to the rules of the auction.
Price quote	Information about the status of the bids currently in the auction.
Bid quote	The amount a seller would have to offer in order to trade.
Ask quote	The amount a buyer would have to offer in order to trade.
Clearance	The process of matching buy and sell bids.
Clearing price	The final transaction price that the buyer pays and the seller receives.

their own profit. They can have different attitudes towards risk: they can be risk prone or risk neutral. The former are likely to raise their bids so that they are more likely to win, whereas the latter tend to bid more conservatively. Some common terminology used in auctions is described in **Table 1**.

The process of an auction usually involves a number of stages. The bidders (buyers and/or sellers) first register with an auction house. Once the auction opens, the bidders offer their bids according to the rules of the particular auction protocol used. A bid indicates a bound on the bidders' willingness to buy or sell a good. Depending on the auction protocol, the auctioneer processes the bids according to the rules of the auction and may generate price quotes accordingly. When the auction closes, buyers and sellers are matched and the transaction price is set. Subsequently, the transactions between buyers and sellers are executed and the buyers pay while sellers ship the goods or provide the service.

Auctions are very versatile protocols and a number of mechanisms have been designed to deal with different situations and requirements.

5.1 Classification of Auctions

Auctions can be characterized along three main dimensions: the *bidding rules*, the *information revelation policy* and the *clearing policy* used.

The bidding rules specify how the auction is to be conducted and what type of bids the participants are allowed to submit:

- Single object or multiple object. In the former, bidders bid on one commodity only, whereas in the latter, bidders are allowed to submit bids for multiple commodities.
- Single attribute or multi-attribute. In single attribute auctions, the negotiation takes place over one dimension or attribute of the good, namely price. In multi-attribute auctions more than one attributes of the good are being negotiated upon, i.e. not only price, but warranty, delivery day etc.
- Single or double. In single auctions, there is a single seller (or buyer) who negotiates with multiple buyers (or sellers), whereas in double auctions there are multiple buyers who negotiate with multiple sellers.
- Open (outcry) or sealed-bid (SB). In open auctions, the bidders bid openly and thus the other participants know theirs bids. In contrast, in sealed-bid auctions the bidders offer their bids in secret, thus not revealing any information to the other participants.
- Ascending or descending price. In the former, the bidders submit increasingly higher bids and the highest bidder gets the item, whereas in the latter the price is lowered by the auctioneer until a buyer decides to bid and obtain the item.
- First-price or second-price (Mth). In first-price auctions, the highest bidder wins and pays the value of her bid, whereas in second-price (Mth) auctions, the highest bidder wins but she only pays the amount designated by the second (Mth) highest bid.

A typology of auctions based on the above distinctions is illustrated in Figure 1. However, this typology is not an exhaustive representation of all different auction formats. In particular, although multi-dimensional auctions such as combinatorial and multi-attribute auctions are included, more

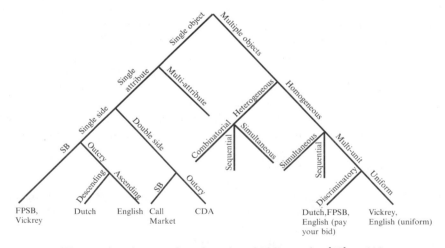

Fig. 1. Auction typology based on bidding rules [10] p. 218

complicated formats such as multi-attribute combinatorial auctions have been left out. Multi-dimensional auctions are in general highly complicated auction formats and are currently the subject of research with somewhat restricted use in practice.

The information revelation policy specifies what information, if any, is revealed to the participants of the auction and has three aspects:

- When to reveal information: on the arrival of each bid, periodically, after a period of activity/inactivity or on market clears.
- What information is to be revealed: if the participants are to obtain information that helps them revise their bids, then the ask and bid quotes need to be made available. Auction closure is another piece of information that can be revealed to the agents as it may be useful in formulating their strategy.
- Who obtains the information: participants only, or everyone.

The clearing policy specifies when the market clears, i.e. when buyers and sellers are matched, and consists of decisions to the following:

- When to clear the market: on arrival of each bid, on closure, periodically, after a period of activity/inactivity.
- Who gets what: who is allocated the goods, how are the winners determined.
- At what prices: what price do buyers pay and sellers receive and how is this determined. Do they pay the first price, second price or some other price. When there are multiple units being auctioned, do all buyers pay the same price (uniform) or a different one (discriminatory).

Finally, auctions can be distinguished into private value, common value and correlated value according to the bidders' valuation of the commodity. In private value auctions, the value of the good depends only on the bidders' own preferences and usually the bidder wants to acquire a good for her own use and consumption. In common value auctions, a bidder's value of an item depends entirely on the others' value of it, which in turn by symmetry is identical to the bidder's, i.e. the good is worth essentially the same to every bidder, but the bidders may have different estimations of this value. In such auctions the bidders want to acquire the commodities for resale or commercial use. In correlated value auctions, a bidder's value depends partly on her own preferences and partly on what she thinks the other agents' valuations are.

5.2 Single Auctions

In single auctions (single good, single attribute), agents are negotiating over one good which is available on its own, the negotiation has one dimension – usually price – and there is one seller (or one buyer) and multiple buyers (or sellers). The most well known basic auction formats are the English, Dutch, First-Price Sealed-Bid (FPSB), and Vickrey.

The English auction is an open-outcry and ascending-price auction which begins with the auctioneer announcing the lowest possible price (which can be the reservation price). Bidders are free to raise their bid and the auction proceeds to successively higher bids. When there are no more raises the winner of the auction is the bidder of the highest bid. The distinct characteristic of the English auction is that the bidders gain information by observing what the others bid. This may be invaluable in forming their own strategy and in some cases even revising their valuation of the auctioned item. There are several variations of the English auction basic format. The bidder's strategy is a series of bids as a function of her private value, her prior estimates of the other bidders' valuations, and the past bids of others. A bidder's best strategy (dominant) is to bid a small amount more than the previous highest bid until she reaches her private value and then stop. In this way a bidder may acquire an object for considerably less than her maximum valuation, simply because she need only increase each bid by a small increment. This on the other hand, means that the seller does not necessarily receive the maximum value for the auctioned good.

The Dutch auction is an open and descending-price auction. The auctioneer announces a high opening bid and then lowers the price until a bidder accepts it. Each bidder needs to decide in advance the maximum amount that she will bid. The bidder must decide when to stop the auction, i.e. submit a bid, based upon her own valuation of the commodity and her prior beliefs about the valuations of the other bidders. No relevant information on the valuation of the other bidders is disclosed during the auction until it is too late. Generally speaking, there is no dominant strategy in the Dutch auction.

In the First-Price Sealed-Bid (FPSB) and Vickrey auctions, the bids submitted are sealed. Sealed-bid auctions have two distinctive phases: the bidding phase in which participants submit their bids, and the resolution phase in which the bids are opened and the winner is determined.

In the FPSB auction the highest bidder wins and pays the amount of her bid. An agent's strategy is her bid as a function of her own private value and prior beliefs about the other bidders' valuations. There is no dominant strategy in the FPSB auction. A high bid raises the probability of winning, but lowers the profit if the bidder is actually awarded the item. Assume that an agent bids her true valuation and it turns out that it is the highest bid, b_h, and wins the auction. Now consider the second highest bid offered b_{h-1}. The winner could have offered just a small increment on that price and still be the winner. Thus the difference between b_h and b_{h-1} represents a loss for the winner. Therefore, agents are better off not bidding their true valuations, but a small amount below it.

In the Vickrey auction, also known as uniform second-price sealed-bid auction, the highest bidder wins but only pays the second highest bid. The agent's strategy is a function of her private value and her prior beliefs of the others' valuations, but the best strategy for a bidder, which constitutes a dominant one, is to bid her true valuation. She then accepts all offers that are below her valuation and none that are above. Truth-telling in the Vickrey auction ensures that globally efficient decisions are being made; bidders do not have to waste time in counterspeculating what the others will do.

In private value auctions, and when the agents are risk-neutral the Dutch and the FPSB auctions are strategically equivalent, i.e. the strategy space is the same, whereas the Vickrey and the English are equivalent, i.e. bidders have the dominant strategy to bid an amount equal to their true valuation. Moreover, assuming valuations are drawn independently, all four mechanisms yield the same revenue on average for the auctioneer (revenue equivalence theorem). If the bidders are risk-prone, then the Dutch and the FPSB auctions yield higher revenue than the English and the Vickrey.

In non-private value auctions, the Dutch is strategically equivalent to the FPSB auction, whereas the Vickrey is not equivalent to the English. The latter is due to the additional information that bidders can obtain about each other's valuations in an open auction. With more than two bidders, the expected revenues are not the same: English>=Vickrey>=Dutch=FPSB (revenue non-equivalence theorem).

5.3 Double Auctions

In double or two side auctions there are multiple sellers and multiple buyers that participate in an auction in order to trade a commodity. These types of auctions are more often used in exchanges and financial markets for trading stocks, bonds, securities, etc. The Continuous Double Auction (CDA) is a general auction mechanism that is used in commodity and stock markets. The general process is as follows:

- Both sellers and bidders submit their bids.
- The bids are then ranked highest to lowest to generate demand and supply profiles.
- From the profiles the maximum quantity exchanged can be determined by matching selling offers with demand bids.
- The transaction price is set and the market clears.

Double auctions may clear either continuously or periodically. In continuous double auctions the buyers and sellers are matched immediately on detection of compatible bids, (stock markets) while in periodic double auctions also known as call markets or clearing houses, bids are collected over specified intervals of time and then the market clears at the expiration of the bidding interval.

One of the issues in double auctions is determining the transaction or clearing price. Consider a set of single unit bids L. M of these bids are sell offers and the remaining $N = L - M$ are buy offers. The Mth price rule sets the clearing price at the Mth highest price among all L bids [35]. The $(M+1)$st price rule sets the clearing price at the $(M+1)$st highest price among all L bids. Determining the bids that are going to transact then, the transaction set, proceeds as follows: while the highest remaining buy bid is greater than or equal to the lowest sell bid, remove these from the set of outstanding bids (breaking the ties arbitrarily) and add them to the set of matched bids (transaction set). For instance, let $M=\{14,13,11,10,5,4,3\}$ and $N=\{12,9,8,7,6,3\}$. The ordered list of all bids $L=\{14,13,12,11,10,9,8,7,6,5,4,3,3\}$. The Mth price is therefore 8 and the $(M+1)$st price is 7. The transaction set consists of the bids $\{(12,3), (9,4), (8,5)\}$. For instance, take the first set of bids in the transaction set (12,3). If the Mth price is used, then the buyer will pay 8 and the seller will receive 8.

The Mth price is undefined if there are no sellers and the $(M+1)$st price is undefined if there are no buyers. In double auctions and as there are multiple buyers and sellers there are mechanisms that allow the agents to obtain information about the current state of the auction in terms of the submitted buy and sell bids. The price quote reveals information to the agents as to whether their bids would be in the transaction set. The bid quote is the price that a seller must offer in order to trade, that is the $(M+1)$st price. The ask quote is the price that the buyer must offer in order to trade, that is the Mth price.

The Mth and $(M+1)$st price rules are generic rules that apply not only in double auctions, but in single auctions as well. In the English auction there is one seller and multiple buyers, thus $M=1$. The seller (auctioneer) may submit a reservation price which can be 0, and according to the rules of the auction the bidders will start submitting successively higher bids. The Mth price in the English auction is going to be the highest bid submitted which determines the winner. In the Dutch auction the seller (auctioneer) starts by announcing a high price (bid) and then successively lowers it until a bidder accepts to pay that price. In this respect, as the auctioneer successively lowers the price, he essentially withdraws the previous bid and submits a new one. M is 1, and this is the Mth price which is the clearing price that the bidder is willing to pay when she announces that she accepts the auctioneer's bid, it is as if the bidder submits a buy bid which is equal to that of the auctioneer. In the first-price sealed-bid auction the Mth price works in exactly the same way as the English auction, the auctioneer may have a reservation price which can be 0, and the Mth price is the highest bid in the auction. In the Vickrey auction, the clearing price is defined as the $(M+1)$st bid among all bids. Again the auctioneer can be perceived as having a bid which is the reservation price or 0, and the bidders submit bids. The winner is the bidder with the highest bid (Mth bid) but she only pays the $(M+1)$st highest bid among all bids.

5.4 Multi-Dimensional Auctions

Buyers and sellers are often interested in other attributes of a good apart from its price. For instance, when one negotiates the purchase of a car with a dealer, she can negotiate not only on the price, but other attributes too, such as for example the warranty, or the extras such a CD player and leather seats or free insurance cover for a year. Auctions that allow bidders to submit bids on more than one attributes or dimensions of a good are called multi-attribute auctions [5]. Such auctions are common in procurement situations [30]. The attributes under negotiation are usually defined in advance, and bidders can compete either in an open-cry or sealed-bid auction on multiple attributes. Mechanism designers for multi-attribute auctions are faced with a number of problems in determining the winner for such auctions, also known as the *winner determination problem*. Multi-attribute auctions enable the application of auction protocols to situations where multiple qualitative attributes need to be taken into account beyond the single dimension of price. This process allows more degrees of freedom for bidders in specifying their bids, while at the same time it allows for an efficient information exchange among the market participants. Such auction formats are currently the subject of extended research and experimentation.

The auction formats described so far allow bidders to bid on individual goods. However, there are situations in which a bidder may not be interested in a single good, but in a variety of distinct, but complementary and interrelated or substitutable commodities. It is thus often the case that goods are only worth to bidders in combination and not when sold separately. So buyers and sellers may have preferences not only for a particular commodity, but for sets or bundles of commodities. By allowing bidders to bid on combinations of different commodities, the efficiency of the auction can be further enhanced. Auctions in which bidders are allowed to bid on bundles of goods are called combinatorial or combinational auctions [8]. Combinatorial auctions are the focus of intense research [8, 23, 24, 25, 28], however, they are currently rare in practice. This has partly to do with the problems regarding efficient and computationally tractable mechanism design for such auctions and partly with the fact that such auctions are cognitively complex and therefore difficult to comprehend by participants.

6 The Need for Market Games

The shift from traditional markets to fully automated electronic ones where various entities such as individuals and organizations are represented by software agents conducting business on their behalf, presents us with a number of challenges. The first challenge that needs to be addressed is that of the infrastructure that is required in place in order for such electronic markets to be fully functional and enable participants first to find and then to interact with each other. The second challenge is with regard to the interactions in such marketplaces. What negotiation protocols or mechanisms are needed to enable participants to interact with each other and reach desirable outcomes. It goes without saying that no unique protocol will suit the needs of all negotiation situations. Each domain has its own characteristics and therefore it is impossible to impose a unique protocol as the panacea for all negotiation situations. The third challenge, is how to build efficient and effective trading agents and endow them with strategies that will enable them to participate and compete in highly complex and dynamic environments that marketplaces are.

Undoubtedly, designing and implementing electronic markets is a complex and intricate process [22]. Agents within such markets are self-interested and are trying to maximize their own utility without necessarily caring about the welfare of others or the society as a whole. As agents are strategic reasoners, if they can gain from lying they will do so, irrespective if this leads to inefficient outcomes in the market. Mechanism design explores such interactions among rational self-interested agents with the view of designing protocols such that when agents use them according to some stability solution concept (dominant strategy equilibrium, Nash equilibrium, mixed strategies Nash equilibrium), then desirable social outcomes follow. But, traditional mechanism design is underpinned by a set of assumptions that may not be realistic for computational agents in complex environments. The most fundamental assumption is that agents are rational and therefore can compute their complete preferences with respect to all possible outcomes. Agents are also assumed to know and understand the protocols and abide by them. In addition, the society of agents participating in the mechanism is static, i.e. the set of agents does not change. Most attention in mechanism design has been focused on centralized mechanisms: agents reveal their preferences to a central mechanism which then computes the optimal solution given these preferences. Furthermore, communication costs are irrelevant in traditional mechanism design and the communication channels are assumed to be faultless. These assumptions are problematic, in particular in the context of electronic markets being populated by software agents, as [7]:

- Agents do not have unlimited memory and computational power.
- Electronic marketplaces are open and unpredictable environments in which agents may cease to operate for a number of reasons.
- In such open systems with heterogeneous agents, a machine-understandable specification of all the associated interaction protocols and associated rules cannot be guaranteed.
- Centralized mechanisms may be unable to compute the outcome because the problem might simply be intractable.
- Communication does not come for free and may not be faultless in a computational setting. Moreover, given the inherent heterogeneity, semantic interoperation between all agents cannot be taken for granted.

Therefore, although traditional mechanism design offers us insights into designing protocols for agent negotiation, its applicability to complex scenarios is inherently limited. The complexity of electronic markets may place them beyond the analytical power of mechanism design. But, even in apparently simple environments, using simple mechanisms or protocols without prior careful design and experimentation, can unfortunately lead to expensive failures or inefficient outcomes, as the conduct of the 3G mobile phone license auctions in some countries in Europe has amply demonstrated [16]. Consequently, developing markets without careful consideration and experimentation can be costly and carries high risks. Moreover, testing trading agents and strategies in real life complex markets is difficult, impractical and also carries high risks. Individuals, businesses and organizations need inexpensive and safe ways to evaluate the appropriateness and applicability of protocols in particular situations as well as the effectiveness and robustness of strategies.

One approach to address these problems is to implement and experiment with market simulations. The intrinsic value of simulations in other domains such as medicine, military applications and education is well-established and accepted. Market simulations offer a number of advantages. Firstly, they allow various parties to test market mechanisms and strategies in a safe environment. For instance, an organization which considers creating an electronic marketplace and using a particular auction as the negotiation protocol of choice can test its appropriateness in this particular domain by experimenting through simulation. The behaviour of participants and their strategies in the marketplace can also be simulated. The advantages and vulnerabilities of strategies can be assessed as well as the impact of using different strategies in the marketplace as a whole. The effectiveness of strategies in different market settings can also be explored. Secondly, using market simulations is also a relative inexpensive means of testing protocols and strategies. When one considers the potential cost of using inappropriate strategies in a marketplace or a negotiation protocol that may be easy to manipulate, experimenting and testing using simulations can significantly reduce the risk of expensive failures. Thirdly, employing market simulations enables the testing of different market settings and configurations and the testing of even extreme conditions and how these affect both the marketplace as a whole as well as individual participants. Extreme even malicious behaviours can also be simulated. In essence, simulations offer a powerful tool which together with formal analysis can provide us with valuable insights into the workings of market infrastructures, the use of protocols as well as the effectiveness of strategies.

Simulated markets may offer the only way to actually approach these problems in a systematic way. Guided first by theory and following a detailed analysis of a given domain, one can proceed to choose appropriate negotiation protocols and then design and implement a market simulation, and finally verify the appropriateness of the protocols chosen through experimentation. The same applies to negotiation strategies.

However, there is one inherent limitation in developing and experimenting with simulations on one's own. The results of such experiments conducted by a single person or organization may lack the touch of realism, since the bidding strategies explored may not necessarily reflect diversity in reality. This is the central idea and major motivation behind the International Trading Agent Competition [30] which features artificial trading agents competing against each other in market-based scenarios. Currently two scenarios are available and run as part of the competition: the Supply Chain Management game and the Market Design game. Ideally, we would like open-source platforms where market-based scenarios could be implemented and researchers have the opportunity to build and test their trading agents against those of others. Such platforms or tools that would allow researchers and developers to design and implement their own marketplaces are currently lacking.

7 Multi-Agent Trading Platforms

One of the first electronic marketplace systems developed was Kasbah [6] in which users were able to create buyer and seller agents that could trade goods on their behalf. When creating a new buying (selling) agent in Kasbah, the user had to define a set of parameters that guided the agent's behaviour. These parameters included:

- desired date to buy (sell) the good by;
- desired price, i.e. the price that the user would like to buy (sell) for the good;
- highest (lowest) acceptable price, which represents the highest (lowest) price that the user is willing to pay (accept) for the good.

The user controlled the agent's behaviour by specifying its "strategy" which was described by the type of price-raise or price-decay function for buyer and seller agents respectively. A price-raise (price-decay) function specified how the agent would raise (lower) its price over time and there were three functions available: linear, quadratic and cubic. Once agents were created, they posted offers in the marketplace and then they waited for other interested parties to contact them. The role of the marketplace was to facilitate connections among agents. For instance, once an agent posted an advert for selling a good, the marketplace returned a list of buyer agents that were interested in the same good as well as notified the respective buyers that a new seller had entered the market. Subsequently, interested parties communicated and negotiated for deals directly. Users had the final say in authorizing the deals that their agents had reached after negotiation. Although users could create new buyer and seller agents, the functionality and strategy of these were completely determined by Kasbah, and users were not able to implement their own agents and plug them into the marketplace.

Another prototype electronic marketplace was MAGMA [32]. The idea behind MAGMA was that of providing the necessary infrastructure and services for building virtual marketplaces. Such services included banking and secure payments, mechanisms for advertising goods as well as transportation of goods. The MAGMA marketplace consisted of the following: trader agents, an advertising server, a relay server and a bank. The trader agents' role was to engage in negotiations to buy and sell goods and services. The advertising server provided advertising and retrieval services while the relay server facilitated communication between the different components in the system. The bank provided basic banking services. The negotiation mechanism used in MAGMA was the Vickrey auction. Unlike Kasbah, agents could be written in potentially different languages as long as they complied with the MAGMA API.

The AuctionBot system was implemented at the University of Michigan as a tool to support research into auctions and mechanism design [34]. AuctionBot was a flexible, scalable, robust auction server capable of supporting both human and software agents. The AuctionBot system had two interfaces: a Web and a TCP/IP interface. The Web interface allowed human agents to register, create, monitor and participate in auctions via web forms. The TCP/IP interface allowed software agents to connect and participate in auctions. The database was used to store the bids submitted by both human and artificial agents. A scheduler continually monitored the database for auctions that had events to process or bids to verify, operating in essence as a daemon process. An auctioneer process would load the auction parameters and the set of current bids from the database and would then validate bids as necessary and could clear the auction or provide price quotes.

A number of auctions were implemented and the system was capable of running many auctions simultaneously. These auctions could be parameterized offering great flexibility. Participation in auctions for trading discrete goods could be on the following basis: {1 seller: many buyers}, {many sellers: 1 buyer}, {many buyers: many sellers}. The system revealed information by providing bid and ask quotes during auctions, and information about past transactions and auctions could also be made available. The auctions' closing time and the timing of clear and/or quote events were among the parameters that could be determined. The Mth and $(M+1)$st price rules were used as the basis for all auctions. In particular the implementation of the Mth and $(M+1)$st rules was based on the 4-heap algorithm [33]. Agents could be written in different programming languages.

The Trading Agent Competition (TAC) [31] is a non-profitable organization whose aim is to promote research into trading agents and electronic markets. The purpose of the competition is to stimulate research in trading agents and market mechanisms by providing a platform for agents competing in well-defined market games with an emphasis on developing successful strategies for maximizing profit in constrained environments. The competition has been running since 2000 and for the first three years it was based on the AuctionBot infrastructure. Since 2003 the underlying infrastructure for the games and the competition has been provided by a team of researchers at the Swedish Institute of Computer Science. The first TAC game was the Travel Agent Game. The market-based scenario featured in TAC travel was that of a travel agent that had to assemble travel packages for a number of clients. Although constructing an agent to take part in an auction for a single good is relatively simple, developing an agent to participate in simultaneous auctions offering complementary and substitutable goods is a complex task. This was the form of the problem that agents had to face in the TAC Travel game. This game was discontinued after 2006. An alternative implementation of the TAC Travel Agent game was provided within the Agentcities initiative which uses a distributed peer-to-peer approach based on FIPA standard infrastructure components, protocols and languages [36].

Currently, TAC operates two games: the supply chain management game (SCM) and the more recent, introduced in 2007, Market Design Game. The first game simulates a supply chain management environment. Traditional SCM deals with the activities within an organization that range from procuring raw materials, manufacturing, to negotiating with customers and acquiring orders from them, and delivering finished products. In today's highly interconnected and networked world, more and more businesses and organizations choose to do business online. This is a dynamic environment where manufacturers may negotiate with suppliers on the one hand, while at the same time compete for customer orders and have to arrange their production schedule and delivery so that customer orders are delivered on time. The ability to respond to changes as they occur and adapt to variations in customer demand and the restrictions as imposed by procurement is of paramount importance. This is the kind of environment that agent technology is best suited for. The TAC SCM game was designed to capture many of the dynamics of such an environment. In the second game, agents are in essence brokers whose goal is to attract potential buyers and sellers as customers and then match these. The interested reader can find more details about TAC and the market games at [31]. Although to some extent the TAC servers that implement the various games are configurable, this is limited to changing certain parameters and there are no provisions for developing games in a systematic way.

One of the first experimental agent-based financial market frameworks was the Santa Fe Artificial Stock Market Model [1, 18].

An alternative approach to developing e-commerce simulation games which builds on and extends the Zeus framework was proposed in [14].

8 e-Game

The **e**lectronic **G**eneric **a**uction **m**arketplace (e-Game) is a configurable auction server that enables both human and artificial agents to participate in electronic auctions [12]. The platform can be extended to support many types of auctions. e-Game has also been designed to support market simulations analogous to those of TAC which can be designed and developed by third parties.

The platform has been developed in JAVA in order to gain the benefits of platform independence and transparency in networking communication and object handling. e-Game is based on a modular architecture separating the interface layer from the data processing modules and the database (Figure 2). The database is used to model the auction space (auction type, startup, closing

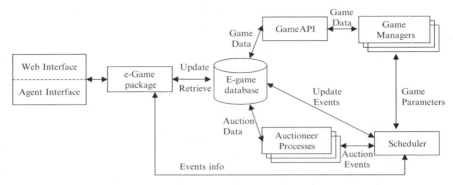

Fig. 2. The architecture of e-Game

time, clearing rules), as well as the user data (registration, user bids, auctions the user has created). It also stores information regarding the market simulations (which game started which auctions, ids of players, scores). It provides fault tolerance, in that the events queue can be reconstructed if there is a problem with the *Auctioneers* or the *Scheduler*. Its modularity and expandability are further strengthened by adopting the object-oriented paradigm for its components.

8.1 Interfaces

e-Game is designed to enable both web users and software agents to participate in electronic auctions. The interface to the outside world consists of two components: the *Web* and the *Agent* interfaces, similarly to [34].

The *Web* interface enables human users to interact with e-Game and provides a series of functions such as registration, creation of new auctions using a variety of parameters (auction type, information revelation, clearing and closing mechanism), bid submission, search facilities according to a series of attributes (auctioned item, status of auction, user's participation in an auction) and viewing the user's profile (change personal information, view previous auction information).

The *Agent* interface enables software agents to interact with the platform and provides agent developers the same functionality with the *Web* interface with respect to auctions plus an additional set of commands for participating in market games. Agents connect to the *Agent* Interface using the TCP protocol and submit their commands using the FIPA ACL language [13]. A subset of the FIPA ACL performatives have been implemented in e-Game to make requests (*request*: when an agent submits a bid), communicate the results of actions (*inform*: when e-Game informs an agent about the result of its bid submission) and refuse requests (*refuse*: when an invalid action has been requested). The actual content of the message follows the FIPA Semantics

Language, which is formally defined in [13]. Nevertheless, a different content-language can be used, for example XML, by developing the appropriate parser and plugging it in, in the existing system. Adopting the FIPA ACL language contributes towards common standards among different agent platforms – if agent technology is to become widely adopted, it is necessary for agents (their developers) to adopt a common communication protocol. For every command an agent submits, e-Game returns an appropriate result together with a command status value that indicates the outcome of the command.

As the majority of the functions that the web users and agents perform are identical, they are implemented in a single package which is shared between the two interfaces. This contributes towards extensibility and also makes the maintenance of the system easier.

8.2 Scheduler

The *Scheduler* runs in parallel with the *Web* and the *Agent* interfaces and is responsible for starting up auctions, games and passing bids to the appropriate *Auctioneer* processes. The auctions can be scheduled by web users or by a specific *GameManager* handler according to a market game's rules, whereas games are scheduled by web users. The *Scheduler* provides a fault-tolerant behaviour, since if it is brought off-line for some reason, when it is restarted it can reconstruct its lists of events by looking into the database for pending events.

8.3 Auction Support

Currently e-Game supports a wide range of auction types (**Table 2**). These basic types of auctions can be further refined by allowing the user to define additional parameters that include [35]:

– Price quote calculation: Upon arrival of new bids, at fixed periods or after a certain period of buy/sell inactivity.
– Auction closure: At a designated date and time or after a period of buy/sell inactivity.

Table 2. Auctions supported by e-Game

Auction type	Sellers	Buyers	Units
English	1	Many	1
Vickrey	1	Many	1
FPSB	1	Many	1
Dutch	1	Many	1
Continuous Single Seller	1	Many	Many
Double	Many	Many	Many

- Intermediate clearing: Upon arrival of new bids, at fixed periods or after a certain period of buy/sell inactivity.
- Information revelation: Whether the price quotes and the closing time are revealed or not.

Such a parameterization can have an effect on the users' or agents' behaviour, since they modify the basic functionality of an auction as well as change the amount of information revealed regarding bids and their status.

Auctions are controlled by the corresponding *Auctioneer* classes. Each *Auctioneer* is started by the *Scheduler* and implements a specific auction protocol. Therefore, there are as many *Auctioneer* classes, as the number of auction protocols supported by e-Game. *Auctioneers* inherit the basic behaviour and state from a *BasicAuctioneer* class. Once an auction starts, the *BasicAuctioneer* class is responsible for generating the auction's events based on the choices the user made during the setup phase. There may be a number of similar or different classes of active *Auctioneers* within the e-Game platform, each one handling an auction with a unique id at any one time.

The implementation of all auctions is based on the Mth and $(M+1)$st price clearing rules. The clearing rules for the auctions are just special cases of the application of the Mth and $(M+1)$st clearing rules as described in section 5.3. However, much depends on the initial set up performed by the user. For instance, to achieve a chronological order clearing in an auction one can set up the intermediate clearing periods to take place as soon as a bid is received and thus the parameterized auctioneer will attempt to perform intermediate clearings upon the arrival of a new bid. As in classical chronological matching, if a portion of the bid cannot transact, then it remains as a standing offer to buy (or sell). Hence, another important aspect of auction implementation has to do with the processing of bids and in particular the states that a bid can be in. The transition diagram of Figure 3 illustrates bid state information in e-Game. The implementation of the Mth and $(M+1)$st clearing rules uses ordered lists.

8.4 Game Management

Game management is facilitated through the *GameManager* class which provides the functionality for the different installed games. There are as many different *GameManagers* as the number of installed games. In the majority of the cases a *GameManager* needs to provide the functionality for the following phases of a game:

1 The initialization phase in which initial resources and future auction events are generated. The resources are allocated to the agents via the TCPServer and the auction events are placed in the *Scheduler*'s events list.
2 The control phase allows one to control certain aspects of the game in an agent-like manner. This means that the developer can implement certain behaviour in their *GameManager* serving the scenario's purposes.

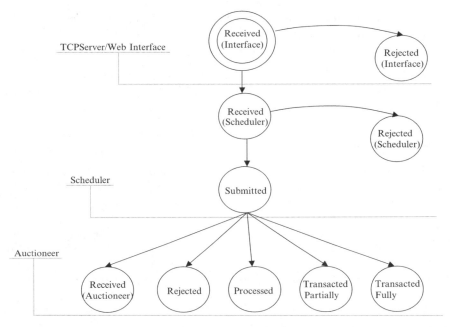

Fig. 3. Bid state transition diagram

For example, it may be appropriate to ensure a ceiling in the prices of some auctions, or buy certain goods for which agents show no preference or trigger events when certain conditions arise in a game.
3 The closing phase, during which the *GameManager* calculates the scores of the participating agents and stores the results in the database for future reference.

9 Developing Trading Games

Apart from the support for agents and web users regarding auction operations, e-Game's main feature is that it supports the design, development and execution of different market simulations or games that involve auctions analogous to those of the Trading Agent Competition.

Market simulations can be distinguished into three categories [14]:

1 Primitive markets in which buyers and sellers interact directly with each other. There are no intermediaries in such markets.
2 Facilitated markets in which there are middle agents such as auctioneers, banks etc. present in the market.
3 Regulated and embedded markets in which there are regulator agents that attempt to control the various interactions among agents having in mind

the social welfare and not only the welfare of individual agents. There may be agents that vary some variables such as availability of goods in the market and interest rates such that they affect the behaviour of others.

We are more interested in the latter two types of market simulations as these can be used to emulate real life complex situations of strategic interdependence among multiple agents.

9.1 Desirable Features

Trading games as other types of games need to be designed carefully in order to be successful, engaging and provide a challenge to the participating players. In particular, when designing a new market game one should have in mind the following key points:

Realism. The more realistic the market simulation is, the more comprehensible it will be to the people who wish to develop agents to participate in it. Ideally, the auction-based market scenario should describe a realistic situation, which may be encountered as part of everyday life (holiday planning, buying interrelated goods, scheduling of resources).

Strategic challenge. The game should present difficult strategic challenges that artificial trading agents may encounter and which require complex decision making. It is up to the agent developers to incorporate and experiment with computational intelligence techniques in their agents, but it is up to the game developer to provide challenges as part of the game.

Efficient use of e-Game. Ideally, the game should not be based on a single type of auction: there is range of auctions to choose from which can be further parameterized. Using different types of auctions provides more challenges for the participants who will have to lay out different strategies depending on the auction type and the amount of information that is revealed at run time.

Fairness with regards to the game specification. In a market game different agents with most probably different strategies will try to accomplish a certain goal. Though the goal will be similar in concept for every agent (for example, assemble one computer by combining different resources), the achievement of individual goals will give different utilities to each agent. One should consider specifications that give everyone the opportunity to come up with the maximum utility.

Fairness with regards to agent design. Agents with a more appropriate bidding strategy should manage to achieve a better outcome, than agents with "naïve" and "simplistic" strategies. When "naïve" strategies actually represent corresponding actions in real life (for which people can actually reason), this prerequisite does not hold. However, the purpose of introducing this key point is that there should not be any flaws in the market game that would allow non-rational strategies to outperform rational ones.

Computational efficiency. The *GameManager* of a market game should have the ability to operate in a real-time manner. Normally, there should not be any algorithms that take a long time to run, especially at critical points in the game. It is acceptable to have such algorithms at the end of the game when an optimal allocation may be required and scores have to be calculated.

Communication efficiency. As little communication as possible should be required in order to reach a desirable outcome.

It is the developer's responsibility to ensure that a game is realistic, fair and non-trivial. Transparency in e-Game is enforced by publishing the logfiles and outcomes of auctions.

Since the e-Game platform allows for web users as well as agents to participate in auctions, it is feasible for everyone to participate in a market game as well. This may sometimes be desirable when comparisons between humans and software agents need to be drawn with respect to laying out a strategy to maximize utility. Moreover, simulations can be run with human agents in order to test selected predications of economic and game theory and psychology and the effects of particular mechanisms and conditions in the market on participants and their strategies.

9.2 Game Design and Development

When designing a market game one often starts with some basic idea of a strategic situation, perhaps inspired by real life and then the details of the game are added in and clarified progressively. The design of the game may have to go through several stages of refinement until all the details are fleshed out.

One of the first issues that needs to be clarified is the objective of the game. In other words, what is it that participants are trying to achieve in the market game and how success is measured. The next issue to be considered concerns the goods, services or other resources that are available in the market and which of these are initially owned by the agents and which are those that can be obtained or traded. The availability of these goods or resources in the market has to be limited so as to encourage competition among agents. This may have to be considered in relation to the number of players participating in a game. The mechanisms that enable the exchange of goods and services in the game need to be decided next. This aspect of game design needs particular attention and obviously needs to be underpinned by theoretical results in mechanism design. For instance, in this stage, decisions regarding what types of auctions will be used and how much information will be revealed to the participants and at which point in the game need to be taken. Deciding if auction closure will be revealed to the participants is an important issue, as in electronic auctions this may result in sniping, i.e. bidders may delay the submission of their bids until the auction is about to close. Bidders may resort to sniping in order to avoid revealing their preferences or their identities early in the auction or avoid driving the prices high quickly, in the hope that ultimately they will get the good at a low price. Furthermore, it is important to consider how competition is induced through the use of different auction protocols and how these affect the participants' behaviour in the game. The game duration needs to be given consideration as well. Relatively simple games may last a few minutes, whereas more complex market games can last longer. In any case, a complex game should not last longer than 60–70 minutes so as to enable participants to play a substantial number of games against various other players to allow them to draw conclusions on their approaches and strategies. Complex games may simulate the passing of time in terms of days and during each day a sequence of activities may have to be carried out by the agents. For instance, information on the availability of goods may be posted in the beginning of the day, while negotiations may follow, and transactions may have to go through by the end of the simulated day.

In terms of implementation, *GenericGame* needs to be extended by the game developers who will have to implement or override a series of abstract methods which are related to generating the each time parameters for the agents that participate in the game, implement the rules of the game, monitor its progress and finally calculate and generate scores. In general, a game involves the following phases:

1   Set up. In this initial phase of the game the developer sets up the game environment and specifies the auctions that are to be run and their parameters.
2   Generation of parameters. These may include client preferences, initial endowments, other information regarding the state of the market, scheduled auctions, availability of goods and other resources and any other data the game developer wants to communicate to each agent in the beginning of the game. The agents are responsible for retrieving this information and using it.
3   Execution. e-Game runs the actual auctions that are part of the game and may also simulate other entities in the market according to the *GameManager* controlling the game. Agents are allowed to submit bids according to the rules of the game. Information is revealed according to the parameters specified in the setup. The agents are responsible for retrieving information on the status of their bids and using this in their decision-making process.
4   Score calculation. Success is measured in terms of a utility function and although this depends on the strategy that an agent is using, the utility function itself needs to be provided by the game developer.

Each game may involve a number of players/agents. The developer may also consider providing a sample agent that the participants can use as the basis to build on. This can also act as a "dummy" to fill in one or more of the available player slots when a game runs and not enough agents have registered for playing. Agent developers will want to be able to follow the progress of their agents in a game. Therefore, a game developer should consider providing the means to facilitate this. Along with the respective *GameManager*, a game developer may also provide an applet to be loaded at runtime which graphically represents progress during the game, so that participants and also others can follow it.

In order to integrate a new game into the e-Game platform one needs to provide the class file, together with an XML file that describes general properties of the game such as the name of the game, the name of the implemented classes (game, applet), a short description for the game, its duration in seconds, the number of allowed participants and the resources (items) that this game uses. This information is then parsed by the *GameImporter* application and is stored in the database. Web users can then browse the web site and following the link *Agent Games* they can view all the installed types of games, together with appropriate links to schedule new instances, watch a current game and view previous scores. At the end of a game, participants can view scores and resources obtained by each agent. When a new game is scheduled, the *Scheduler* receives the corresponding event and at the appropriate time loads the user defined *GameManager* class.

9.3 The Computer Market Game

To provide the reader with an idea of the type of market simulations that can be created using e-Game we will briefly discuss one such market simulation[1]. In the Computer Market Game (CMG), which lasts nine minutes, each of the six agents participating is a supplier whose task is to assemble PCs for its five clients.

There are only three types of parts that are necessary to make up a properly working PC: a motherboard, a case and a monitor. There are three types of motherboards, each one bundled with CPUs of 1.0 GHz (MB1), 1.5 GHz (MB2) and 2.0 GHz (MB3). There are two types of cases, one with a DVD player (C1) and the other with a DVD/RW drive (C2). There is only one type of monitor.

In the beginning of the game the agents receive their clients' preferences in terms of a bonus value for upgrading to a better motherboard (MB2 or MB3) and a better case (C2). An example set of preferences for an agent A is given in **Table 3.** Hence, client 2 of agent A offers 115 monetary units for motherboard MB2 and 175 for the MB3 one. For obtaining the better case (C2) the client offers 290 monetary units. e-Game generates values in the following ranges for these goods: MB2 = [100...150], MB3=[150...200], C2=[200...300].

Components are available in certain quantities and are traded in different auctions (**Table 4**). Figure 4 illustrates how the auctions for the various items are scheduled during the 9-minute period of the game. The dotted part of the lines indicates that the auctions may close anytime during that period, but the exact closing time is not revealed to the agents.

---

[1] Note: this is a simple market simulation which has been developed for teaching and not research purposes.

Table 3. An example of client preferences

Agent	Client	MB2	MB3	C2
A	1	120	165	234
A	2	115	175	290
A	3	140	190	219
A	4	145	187	270
A	5	135	164	245

Table 4. Availability of goods and auctions in the CMG game

Component	Quantity	Auction
MB1	17	Mth Price
MB2	8	Mth Price
MB3	5	Mth Price
C1	20	Mth Price
C2	10	Mth Price
M	30	Continuous Single Seller

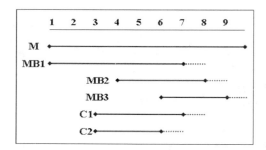

Fig. 4. The auctions' schedule in the CMG game

An agent's success in this game depends on the satisfaction of its clients. An individual client i's utility (CU_i) is:

$$CU_i = 1000 + MotherboardBonus + CaseBonus \qquad (1)$$

For each client that has been allocated a fully-assembled PC, the agent gets 1000 monetary units plus any bonus for upgrading to a better motherboard or case. If no fully-assembled PC is allocated to a client, then this utility is 0. For every extra component purchased which exceeds the quantity needed to satisfy its clients the agent has to pay a penalty which is perceived as a storage cost and is determined at the beginning of the game as a random amount between 150 and 300 units. This has been added as a disincentive to buy extra items surplus to requirements. An agent's utility function is the sum of all the individual client utilities minus the expenses (*Expenses*) and the penalties (*Penalties*) incurred:

$$AU = \Sigma(CU_i) - Expenses - Penalties \qquad (2)$$

The agent's strategy should be focused on providing a fully-assembled PC to each one of its clients or to as many as possible, while at the same time trying to minimize costs. There are obvious interdependencies between goods, as a fully-assembled PC requires three components. In creating a strategy for this game, one has to take into account that the availability of goods is limited, prices in auctions may vary, auctions close at different times and therefore one may have to switch to a different auction if they fail to acquire a certain good, and customers give different bonuses for upgrading to a better specification. Moreover, an agent's success does not only depend on its own strategy, but crucially that of the other agents too. Nevertheless, an agent does not have any means of knowing the identity of the other players or monitoring their bids directly.

The original version of the CMG game did not include any penalties. In the process of using the game with a group of students who had to develop agents as part of their coursework, we discovered that a significant number of agents were buying more than they needed in an effort to deprive the market of various goods and therefore lower the utility of the other participants. To this end, we extended the agent's utility function to account for penalties in the form of storage costs for extra components. This is an example where experimentation has revealed a flaw in the design of the market game and which had to be rectified.

In order to implement the above scenario the developer would have to provide an XML file describing the general properties of the game (names of game and applet classes, number of players, duration of the game, auctioned resources), together with the appropriate classes. The specific *GameManager* would schedule the auctions at the beginning of the game by implementing the method *startupGame* and using the *scheduleAuction* methods. The next step would be to generate the parameters that each agent receives by implementing the method *generateParameters*. The utility of each agent would be calculated at the end of the game by examining the winning bids of each auction. Finally, any language resources that the developer used would be freed-up in the *closeGame* method. In addition, the developer may choose to write an applet so that web users can view the progress of their agent. Such an applet for this simple game is illustrated in Figure 5.

A number of other games have also been developed based on the e-Game infrastructure by us but also by students. All these games are publicly accessible from e-Game's website[2].

---

[2] http://csres43:8080/egame/index.jsp and http://sh718:8080/egame/index.jsp

Fig. 5. The CMG applet

10 Further Work

e-Game can be further extended in a number of ways. A possible direction is with regards to multi-attribute auctions. In a multi-attribute auction, the auctioneer lists a number of negotiable attributes and bidders are asked to submit multi-dimensional bids. Multi-attribute auctions are often used in selecting suppliers for various services in the public sector. The heterogeneity of the negotiable attributes often calls for more consideration during the bidding phase, since bidders do not only compete on price, but also on other features such as quality, additional features on the requested item, availability, extended warranty, provision of additional resources and more. The inclusion of such a feature increases the complexity of the implemented marketplaces, since evaluating heterogeneous bids is a complex problem. There have been works that suggest ways to maximize the auctioneer's turnover [5]. This extension also contributes to increasing the complexity on the bidder's side (game participants) that now need to consider a number of attributes when submitting their bids and therefore provides more opportunities to agent developers to include computational intelligence techniques in their agents. The inclusion of such a feature is non-trivial, but it will provide insight into the topic of multi-attribute auctions.

In a fully automated marketplace, agents are not only concerned with the process of negotiation, but also proactively seek opportunities to trade. To enable agents to search for scheduled auctions or even set-up auctions, the *Agent* interface can be extended in order to provide the same functionality to software agents that the *Web* interface provides to humans. We are also considering allowing agents to initiate auctions during a game and act as sellers of goods that they own, or have previously acquired through trading, but no longer desire or they are surplus to requirements.

Although e-Game's components were developed using Java, a web services interface can also be built that would allow game developers to have access to e-Game's API using the programming language of their choice.

Finally, we are considering the possibility of introducing a rule-based specification such as [19] in order to describe complex market-based scenarios.

11 Conclusions

Although agents and multi-agent systems are regarded by many practitioners, researchers and technologist analysts as the technology that has the potential to revolutionize electronic commerce, so far their immense potential has not been fully realized. This chapter discussed how some of the obstacles of deploying agent technology in e-commerce applications can be overcome by developing electronic market games, that can be used to conduct research on market infrastructure, negotiation protocols and strategies.

e-Game is a generic auction platform that allows web users as well as agents to participate in auctions and auction-based market games. The aim of the e-Game project is to offer the infrastructure for running complex market simulations and conducting experiments with different bidding strategies and enable agent developers to incorporate and experiment with computational intelligence techniques. The most important feature of e-Game and what distinguishes it as a platform from other efforts such as [32, 6, 34, 31, 15] is that it provides independent developers the facilities to design, implement and run auction-based market simulations. Unlike JASA [15] for instance, e-Game does not simply provide the facilities for scheduling, running or conducting experiments on the use of strategies in standalone auctions. Individual users/developers can write their own *GameManager* and define complete market scenarios together with accompanying applets and sample agents that can be inserted and run on top of the e-Game infrastructure. The development of these modules is completely separate from the rest of the system. By making use of existing classes and implementing simple interfaces, users can develop new games quickly. The progress of integrating new games with e-Game is fully automated: since the user developed classes implement certain (simple) interfaces and inherit from existing classes, e-Game can access these methods without needing to know the rest of the user-defined methods that implement the game's rules.

Moreover, e-Game provides the facility of having humans participate in complex market simulations as well as humans playing against software agents. This may be useful in designing strategies and examining strategy efficiency and complexity. One could study the benefits from using agents in complex markets in a systematic way. For example, in a simple game, a human may do better than an agent, but it would be interesting to pit humans against software agents and study what happens when the scenario gets more complicated (penalty points, allocation problems etc). A suitable version of the CMG game for human users has already been developed and a number of laboratory-based experiments have been conducted.

Apart from using the platform for designing and running complex market games, it can also be used for teaching negotiation protocols, auctions and strategies. Students can have a hands-on experience with a number of negotiation protocols and put into practice the principles taught.

Although the work presented here goes some way towards addressing some of the issues of deploying agents in e-commerce applications, we should not forget one other fundamental issue: users need to trust software agents in order to be able to confidently delegate to them the tasks of negotiating and reaching deals on their behalf and they also need to be assured that any legal issues relating to agents trading electronically are fully covered, just as they are in traditional trading practices [11].

12 Resources

Key books

- Baba, N. and Jain, L.C., Computational Intelligence in Games, Springer, 2001.
- Fasli, M. Agent Technology for E-commerce. John Wiley and Sons, 2007.

Key papers and other resources

- Agorics. Auctions. http://www.agorics.com/Library/auctions.html
- Griss, M. and Letsinger, R. Games at work-agent-mediated e-commerce simulation. In Proceedings of the 4th international conference on Autonomous agents (Agents 00), Barcelona, Spain, 2000.
- Sadeh, N. M., Arunachalam, R., Eriksson, J., Finne, N., and Janson, S., TAC'03: A Supply Chain Trading Competition, AI Magazine, 24(1):92-94, 2003.
- Wellman, M. P., Wurman, P.R., O'Malley, K., Bangera, R., Lin, S.-D., Reeves, D. M., and Walsh, W.E., Designing the Market Game for a Trading Agent Competition, IEEE Internet Computing, 5(2):43-51, 2001

Organizations, Societies, Special Interest Groups, Projects

- IEEE Specialist Interest Group in Computational Intelligence in Games, http://ieee-cs.org/games/
- Trading Agent Competition (TAC) Organization. http://tac.eecs.umich.edu/association.html
- Agentcities Project, http://www.agentcities.org/

Key International Conferences/Workshops

- Agent-Mediated Electronic Commerce Workshop (AMEC)
- Autonomous Agents and Multi-agent Systems Conference (AAMAS)
- IEEE International Conference on Electronic Commerce (IEEE ICEC)
- IEEE Symposium on Computational Intelligence in Games (IEEE CIG)
- Trading Agent Design and Analysis Workshop (TADA)

Software

- electronic Generic Auction Marketplace (e-Game) servers: http://csres43:8080/egame/index.jsp and http://sh718:8080/egame/index.jsp
- Trading Agent Competition Servers: http://www.sics.se/tac/
- TAGA and Agentcities, http://taga.sourceforge.net/doc/agentcities.html

Acknowledgements

The introduction to negotiation protocols and auctions is based on Fasli, M. Agent Technology for e-Commerce, © John Wiley and Sons, 2007. Reproduced with permission.

References

1. Agorics. Auctions. http://www.agorics.com/Library/auctions.html.
2. Arthur, W. B., Holland, J., LeBaron, B., Palmer, R. and Tayler, P. Asset pricing under endogenous expectations in an artificial stock market. In Arthur, W. B., Durlauf, S. and Lane, D., editors, The Economy as an Evolving Complex System II, Addison-Wesley Longman, Reading, MA, pages 15–44, 1997.
3. Bakos, Y. The emerging role of electronic marketplaces on the internet. Communications of the ACM, 41(8):35–42, 1998.
4. Bichler, M. A roadmap to auction-based negotiation protocols for electronic commerce. In Hawaii International Conference on Systems Sciences (HICSS-33), 2000.
5. Bichler, M. and Kalagnanam, J. Configurable offers and winner determination in multi-attribute auctions. European Journal of Operational Research, 160(2):380–394, 2005.
6. Chavez, A. and Maes, P. Kasbah: An agent marketplace for buying and selling goods. In First International Conference on the Practical Application of Intelligent Agents and Multi-Agent Technology (PAAM'96), pages 75–90, 1996.
7. Dash, R. K. and Jennings, N. R. and Parkes, D. C. Computational-mechanism design: A call to arms. IEEE Intelligent Systems, 18(6):40–47, 2003.
8. de Vries, S. and Vohra, R. Combinatorial auctions: A survey. INFORMS Journal on Computing, 15(3):284–309, 2003.
9. Ebay. http://www.ebay.com/.

10. Fasli, M. Agent Technology for e-Commerce. John Wiley and Sons, Chichester, 2007.
11. Fasli, M. On Agent Technology for E-commerce: Trust Security and Legal Issues. Knowledge Engineering Review (in press), 2007.
12. Fasli, M. and Michalakopoulos, M. e-Game: A generic auction platform supporting customizable market games. In Proceedings of the IEEE/WIC/ACM Intelligent Agent Technology Conference (IAT 2004), pages 190–196, Beijing, China, 2004.
13. FIPA Communicative Act Library Specification. http://www.fipa.org/specs/fipa00037/, 2002.
14. Griss, M. and Letsinger, R. Games at work-agent-mediated e-commerce simulation. In Proceedings of the 4th international conference on Autonomous agents (Agents 00), Barcelona, Spain, 2000.
15. JASA: Java Auction Simulator API. http://www.csc.liv.ac.uk/~sphelps/jasa/.
16. Klemperer, P. How (not) to run auctions: The European 3G Telecom Auctions. European Economic Review, 46(4–5):829–845, 2002.
17. Kumar, M. and Feldman, S. Internet auctions. In Proceedings of the 3rd USENIX Workshop on Electronic Commerce, pages 49–60, 1998.
18. LeBaron, B., Arthur, W. B. and Palmer, R. Time series properties of an artificial stock market. Journal of Economic Dynamics and Control, 23:(1487–1516), 1999.
19. Lochner, K. M. and Wellman, M. P. Rule-based specifications of auction mechanisms. In Third International Joint Conference on Autonomous Agents and Multi-agent Systems Conference (AAMAS), pages 818–825, 2004.
20. Maes, P., Guttman, R., and Moukas, A. Agents that buy and sell: Transforming commerce as we know it. Communications of the ACM, 42(3):81–91, 1999.
21. Mas-Colell, A., Whinston, M. D., and Green, J. R. Microeconomic Theory. Oxford University Press, Oxford, 1995.
22. Neumann, D. and Weinhardt, C. Domain-independent enegotiation design: Prospects, methods, and challenges. In 13th International Workshop on Database and Expert Systems Applications (DEXA'02), pages 680–686, 2002.
23. Nisan, N. Algorithms for selfish agents. In Proceedings of the 16th Annual Symposium on Theoretical Aspects of Computer Science, (STACS'99), LNCS Volume 1563, pages 1–15. Springer, Berlin, 1999.
24. Nisan, N. Bidding and allocation in combinatorial auctions. In Proceedings of the 2nd ACM Conference on Electronic Commerce (EC-00), pages 1–12, Minneapolis, MN, 2000.
25. Nisan, N. and Ronen, A. Algorithmic mechanism design. Games and Economic Behavior, 35:166–196, 2001.
26. onSale. http://www.onsale.com/, 2005.
27. Parkes, D. C. Iterative Combinatorial Auctions: Achieving Economic and Computational Efficiency. PhD thesis, University of Pennsylvania, 2001.
28. Pekec, A. and Rothkopf, M. H. Combinatorial auction design. Management Science, 49(11):1485–1503, 2003.
29. Sandholm, T. Distributed rational decision making. In G. Weiss, editor, Multi-agent Systems: A Modern Approach to Distributed Artificial Intelligence, pages 201–258. The MIT Press, Cambridge, MA, 1999.
30. Sun, J. and Sadeh, N. M. Coordinating multi-attribute procurement auctions subject to finite capacity considerations. Technical Report CMU-ISRI-03-105,

e-Supply Chain Management Laboratory, Institute for Software Research International, School of Computer Science, Carnegie Mellon University, 2004.
31. Trading Agent Competition. http://www.sics.se/tac/.
32. Tsvetovatyy, M., Gini, M., Mobasher, B., and Wieckowski, Z. Magma: An agent-based virtual market for electronic commerce. Journal of Applied Artificial Intelligence, 6, 1997.
33. Wurman, P., Walsh, W. E., and Wellman, M. P. Flexible double auctions for electronic commerce: theory and implementation. Decision Support Systems, 24:17–27, 1998.
34. Wurman, P., Wellman, M. P., and Walsh, W. The Michigan Internet Auction-Bot: A configurable auction server for human and software agents. In Proceedings of the Autonomous Agents Conference, pages 301–308, 1998.
35. Wurman, P. R., Wellman, M. P., and Walsh, W. E. A parameterization of the auction design space. Games and Economic Behavior, 35(1):304–338, 2001.
36. Zou, Y., Finin, T., Ding, L., Chen, H., and Pan R. TAGA: Trading agent competition in Agentcities. In Workshop on Trading Agent Design and Analysis held in conjunction with the Eighteenth International Joint Conference on Artificial Intelligence (IJCAI), 2003.

EVE's Entropy: A Formal Gauge of Fun in Games

Kevin Burns

The MITRE Corporation, 202 Burlington Road, Bedford, MA, 01730-1420, USA
kburns@mitre.org

Abstract. Fun is the force that drives play in games. Therefore a science of fun is needed to engineer games that are engaging for players, including the design of software agents that play against and along with human beings in computer games. EVE' is a computational theory, based on a psychological progression of Expectations (E), Violations (V) and Explanations (E') that in turn evoke emotional responses known as *fun* or *flow*. Here EVE' is developed for the case of gambling games, specifically slot machines, deriving a Bayesian-information measure of aesthetic utility that is not modeled by the economic utility of previous proposals like Prospect Theory. The derivation shows how aesthetic utility can be measured by entropy and how fun can be seen as a form of learning. EVE's contribution lies in going beyond classical economics and computational intelligence to analyze the aesthetics of enjoyment and engagement – towards a science of fun.

1 Introduction

People play games for many reasons, but mostly for fun. Yet the field of game design in general and computational intelligence in particular is lacking a *science of fun*.

This science must go beyond computational *intelligence* to address computational *aesthetics*. A science of fun must also go beyond the *computational* logic of software agents to address the *psychological* basis for intelligent thinking and aesthetic feelings in human beings. Simply stated, if artificial agents are to deal well with people then they must be endowed with both intelligence and aesthetics – of the same sorts that govern human cognition and emotions.

Like all science, a science of fun requires test beds in which analyses and experiments can be performed, as well as theories that can guide testing and be generalized to applications. Here the focus is on gambling games, particularly simplified slot machines. Section 2 describes this test bed and highlights the difference between *economic utility* and *aesthetic utility* in slots and other gambles. Section 3 presents a psychological theory and computational model called EVE' [1], which measures fun in games based on Bayesian probabilities and Shannon-like entropies. Section 4 concludes with a look at how EVE′ can be extended and applied to other games and any work – including computer games played for "serious" purposes in military, business and other domains.

2 Gambling Games

2.1 Looking at Slots

Perhaps the first computer-video game, in the sense of a machine that calculated mechanically and interacted visually with human players, was the slot machine. Yet despite the large literature on human-machine interaction in computing media, surprisingly little attention has been paid to slots. Huhtamo [2] provides a socio-historical review, but neither he nor anyone else has (to the best of my knowledge) explored the psycho-computational basis for fun in slots [3].

In fact at first blush it seems rather surprising that a game as simple as slots could engage and entertain intelligent human beings, many thousands of them each day, for so many thousands of days. Perhaps this was less surprising before the days of digital computers, but the continued popularity of casino slot machines raises new questions about the nature of fun in games – especially computer games. That is, while much hype is made of virtual reality and artificial intelligence and other new and high technologies, many people still pay to play slot machines – so much so that slots are arguably the most popular and profitable game played on computing machines today. This suggests that fun in slots can shed light on the fundamental nature of fun in games.

Clearly the fun of slot machines involves *actions* like pulling the arm, as well as *sensations* from coins, bells and lights. But these have only secondary effects on fun, as we know from field observations as well as lab experiments. In the field (casino), slots are often played without pulling an arm (just pressing a button) and without payoffs in coins (just getting a credit). In the lab, many studies (discussed below) show that human choices in gambling vary with an abstract parameter P (probability of a win) that has nothing to do with a physical action or sensation per se. Therefore most fun (at least in gambling) must be mental, stemming from *cognition* (thinking) that leads to *emotions* (feelings). And yet there seems to be little or no thinking when one plays slots, since the game involves just pulling (or pressing) then "see what comes and deal with it".

Below I dissect the nature of this "see what comes and deal with it" experience from a psychological perspective using mathematical modeling. The approach, called EVE′, employs an information-theoretic notion of entropy – and decision-theoretic notions of probability and utility. I begin with probability and utility.

2.2 Taking a Chance

When cut to the core, slot machines are much like any other lottery in which a player gives up a sure-thing (ante A) for a chance to win more *utility* (payoff J). Mathematically, a typical slot machine has a set of N possible payoffs $\{J_1, J_2, ..., J_N\}$, each with a *probability* P of occurrence $\{P_1, P_2, ..., P_N\}$. Typically each J_n is inversely proportional to the associated P_n.

The number N of possible payoffs varies from machine to machine, but any machine can be characterized as paying a surrogate utility $J \approx 1/P$ at an aggregate probability $P = \Sigma_N P_n$. Here to simplify the analysis I will take the aggregate P and surrogate J as parameters of a "test tube" machine that either pays a jackpot (win of J) or nothing (loss of A) on each turn.

The product P*J is called an *expected utility*. When A = P*J the machine is said to be "fair" and in this case the player has no economic incentive to play. For a "real" machine A > P*J so the House can make a profit (typically 5%). This makes slots a losing proposition for the player, from an *economic* perspective – and yet people play! Therefore slots must be a winning proposition from some other perspective, and here I suggest it is an *aesthetic* perspective, i.e., people play because slots are fun.

More formally I propose that players must be getting some positive *aesthetic* utility that equals or exceeds the magnitude of their negative (~5%) *economic* utility, and this aesthetic utility is what we call fun. The challenge for a science of fun is to compute aesthetic utility, which EVE′ does in Section 3. But before presenting EVE′ it is useful to review findings from psychological experiments that are not well explained by existing theories of economic utility, which do not consider aesthetic utility.

Returning to the basic tradeoff between A and P*J, consider the following choice between a certain amount (like A) and possible payoff (like P*J):

You have a choice between: (s) a sure-thing of getting $20; or (g) a gamble with 20% chance of getting $100 and 80% chance of getting nothing.

Note the s and g options are economically equivalent, i.e., you have no economic incentive to choose s (with expected utility of $20) over g (with expected utility of $20) or vice versa. Thus we would expect no preference, such that s and g would each be chosen about half of the time. However, formal testing [4] shows that people prefer the gamble (g) to the sure-thing (s), i.e., people are *risk seeking*. Now consider the following choice:

You have a choice between: (s) a sure-thing of getting $80; or (g) a gamble with 80% chance of getting $100 and 20% chance of getting nothing.

Again, note the s and g options are economically equivalent, i.e., you have no economic incentive to choose s (with expected utility of $80) over g (with expected utility of $80) or vice versa. Thus again we would expect no preference, and again we would be wrong. But this time formal testing [4] shows that people prefer the sure-thing (s) to the gamble (g), i.e., people are *risk averse*.

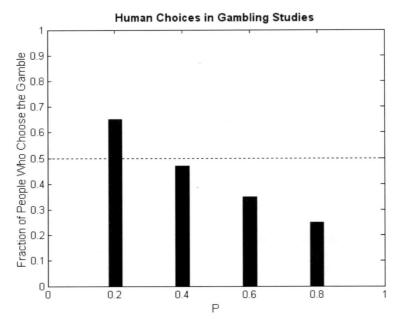

Fig. 1. Fraction of people who choose the gamble over an economically-equivalent sure-thing. Data from [4]. People are risk seeking for P=0.2 and risk averse for P≥0.4

These findings (see Fig. 1) illustrate that, in the absence of economic incentive, human beings exhibit biases (risk seeking and risk averse) that depend (among other things) on the probability (P) of a gamble. Such findings, which have been replicated many times, are a major problem for classical economic theories that assume choices are governed by expected utilities like P*J. In fact the widely-acclaimed *Prospect Theory* [5] was developed precisely to explain these and other anomalies that cannot be explained by previous theories of economic utility.

Prospect Theory postulates three deviations from classical economics in order to explain gambling preferences. One is a "framing effect", in which utility is measured relative to a reference point such that choices will be different for gambles to gain (when J is positive) versus gambles to pay (when J is negative). With respect to this distinction, all gambles in Fig. 1 and slot games are gambles to gain – so a framing effect cannot explain why gambling preferences vary with P in these cases.

Another dimension of Prospect Theory replaces utilities like J and A with non-linear *value functions*, whereby money is assumed to be marginally discounted – such that the subjective value of a dollar won or lost decreases as the number of dollars won or lost is increased. This helps to explain a "framing effect" (discussed above) but at the same time it predicts greater risk aversion for all gambles to gain, since the payoff J is discounted more than the sure-thing A. Therefore the value function of Prospect Theory makes it even harder to explain risk seeking choices like those seen at P=0.2 in Fig. 1, and harder still to explain risk seeking in the case of "real" slots where risk aversion should be greater since A > P*J.

A final dimension of Prospect Theory, designed to account for the risk seeking preference seen at low P (Fig. 1), replaces probabilities like P with a *weighting function* W. This W is an empirical fit to data (see Fig. 2) – and although the mathematical function has desirable features [6] it is questionable in its psychological basis. In particular, the function for W versus P (Fig. 2) fits the data but raises the question: Why an S-shape?

As a preview of EVE′ it is useful to wonder whether the preferences plotted in Figs. 1 and 2 are governed by an *aesthetic* utility, rather than by modification to the P of *economic* utility as in the W of Prospect Theory. Fig. 3 plots what this aesthetic utility would be, if it were backed-out of the weighting function and instead added to economic utility.

Later (Section 3) EVE′ will derive a measure of aesthetic utility as a function of P that looks almost exactly like Fig. 3. But first it is useful to examine other theories of aesthetic utility (fun) proposed before EVE′.

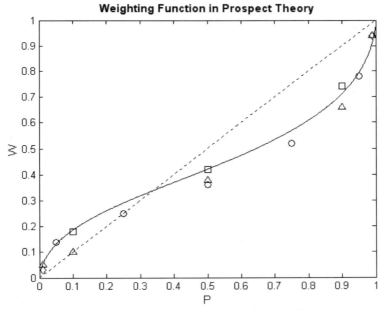

Fig. 2. S-shape of weighting function in Prospect Theory [5], $W = P^\gamma / (P^\gamma + (1-P)^\gamma)^{1/\gamma}$, with $\gamma=0.61$. Data at J=$50 (squares), J=$100 (circles), J=$200 (triangles) and J=$400 (diamonds)

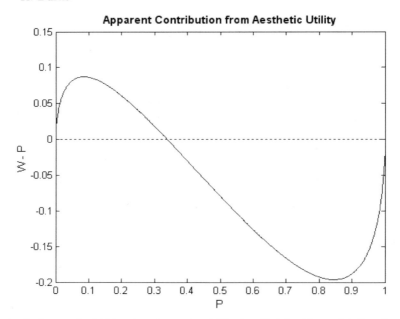

Fig. 3. S-shape of addition to economic utility implied by the weighting function of Prospect Theory (see Fig. 2). This is the apparent contribution from aesthetic utility

2.3 When Will We Learn

Unlike theories of economic utility, which are numerous, formal theories of aesthetic utility are rare [1]. This is due at least in part to the reality that necessities like food, bought with economic utility, are commonly considered to be much more important than niceties like fun, brought by aesthetic utility. Nevertheless, aesthetics are an essential component of the human condition, and in fact some have argued that the emergence of art itself marks a turning point in evolution where the human mind as we know it was created [7].

Moreover, food is fun as well as nourishing so the line between economics and aesthetics is not so clear. This is especially true in our post-industrial information age, where a "rise of aesthetic value" [8] in commerce and culture has fueled a renaissance in aesthetic analysis and artistic synthesis [9] including new media and game design studies on the nature of play [10].

But even in this renaissance, formal theories of "fun" are hard to find. One informal theory, which was proposed by a game design professional [11], is not really a theory in the scientific sense of being specific and testable. However this theory does give some insight into what makes games fun, when it sums up its message as follows (pg. 46): "Fun is just another word for learning."

This is consistent with common sense, since learning to master a task does seem to be part of the fun in a game. But all learning is not so fun, in games or other work, and even if all learning were fun it is not clear how this learning could be quantified – hence how fun could be engineered in *computational aesthetics*. In particular, for the case of slot games, it is not obvious what players are learning and why it is fun.

Below I outline more precisely what it is that players are learning when they play slots. I also discuss why some learning is fun (e.g., on a win) while some learning is not (e.g., on a loss), and how the two combine computationally to affect the aesthetic experience. Some general concepts like learning and winning have been discussed by many others. But a specific and testable theory is missing, and that is where EVE' is aimed.

3 Feeling Fun

This section reviews the atoms of EVE' [1], which are mathematical measures of psychological processes, namely: Expectation (E), Violation (V) and Explanation (E'). EVE' holds that a cognitive progression of E-V-E' is the core of every aesthetic experience, although this progression can be complicated by hierarchical nesting in temporal and other dimensions.

The beauty of "fair slots" (where A = P*J), which are analyzed here, is that the game can be characterized by a single parameter (P) that makes the atoms of EVE' more visible and computable. The beauty of EVE' is that the same atoms, laid bare in slots, are also the building blocks of more complicated aesthetic experiences. This allows a basic understanding of EVE' in slots to be extended and applied to other games and media computing (see Section 4).

The fundamental assumption underlying EVE' is that enjoyment (fun) comes from success in cognitive processing, as a person "makes sense" of a media experience. It feels good to make sense, and the good feeling depends on how hard it is to make sense as well as how much sense is made. This creates a tradeoff between the *pleasure* that comes when *expected* events occur – versus the *pleasure-prime* that comes from *explaining* events that were not expected. These two good feelings are different but related since the latter (Explanations), commonly referred to as learning, can only come at the expense of the former (Expectations).

Between EVE's pleasure from Expectations (E) and pleasure-prime from Explanations (E') lies tension in Violations (V) of Expectations – which require Explanations (E') in order to make sense of what has been seen (or heard, etc.) in the media experience. More formally, these measures (E, V, E') can be modeled by the mathematics of Shannon information theory [12] and Bayesian decision theory [13].

3.1 Pleasure and Pleasure-prime

EVE's enjoyment (fun) comes in two ways. One way is the pleasure (p) that arises when Expectations (E) are formed and these Expectations pan out,

160   K. Burns

i.e., when observations are consistent with Expectations. Another way is the pleasure-prime (p′) that arises after Violations (V) of Expectations (E), when Explanations (E′) are formed to make sense of the Violations. These two pleasures are different, as noted above – and the engineering of aesthetics can be seen as an artist's (creator's) attempt to optimize this E-E′ tradeoff in the mind of an audience (consumer).

EVE's mathematical measures of E and E′, as well as p and p′, are derived and discussed elsewhere [1, 3, 14] and [15]. Here my purpose is only to summarize the equations and illustrate some key features. Beginning with E, EVE′ uses Shannon's [12] "intuitive" information-theoretic measure, log P, as a measure of Expectation for an event expected with probability P. Likewise, EVE′ uses −log P as a measure of Violation for the same event expected with probability P. Finally, EVE′ uses a "posterior" Bayesian-theoretic factor R to measure the extent of *resolution* for an Explanation (E′) of Violation (V = −log P), such that E′ = −log P * R. This R will be discussed further below (see "Believing in Bayes"), but for now the resolution is assumed to be complete such that |R|=1 for a win or loss event in slots.

Focusing on a "test tube" slot game where there are only two outcomes, with probabilities P and Q for win (of jackpot) and loss (of ante), respectively, EVE's equation for E is as follows:

$$E = [P^* \log P + Q^* \log Q]. \qquad (1)$$

This equation is a total measure of success in forming Expectations, where each "atomic" term (log P and log Q) is weighted by the "average" frequency (probability) at which the associated event (win or loss) occurs over a series of turns.

The equation for E′ is similar but reflects two major differences. One difference is that the terms for P and Q each contain a factor R to account for the valence (+ or −) and magnitude of resolution. As discussed below ("Believing in Bayes"), the magnitude of R is equal for win and loss events in slots; |R|=1 in each case, which means that resolution (measured as a fraction of the Violation) is complete. However the difference in valence (positive or negative) for wins and losses introduces a difference in sign (+ or −). This can be contrasted with E, where pleasure arises from success in forming Expectations so the sign is positive for *any* event (good or bad).

Another feature of E′ is that the magnitude of pleasure-prime arising from a unit of resolution (R), when a person "gets it", will depend on the person's *sense of humor*. That is, a unit of good resolution will produce stronger feelings than a unit of bad resolution when a player has a good sense of slot humor, and vice versa when a player has a bad sense of slot humor. Thus using H^+ (for win) and H^- (for loss) as weighting factors to model a player's sense of slot humor, and assuming the resolution of Violation is always complete such that |R|=1, EVE's equation for E′ is written as follows:

$$E' = -[H^{+*}P^* \log P - H^{-*}Q^* \log Q]. \qquad (2)$$

Note that besides R and H, including the difference in sign associated with good (win) and bad (loss) outcomes, this equation for E′ differs from that of E by an overall minus sign. Thus the term containing H^+ is negative and the term containing H^- is positive, because E′ is measured as a fraction (R) of a Violation (V) and V is the negative of E (see above).

Note also that these equations for E and E′ contain marginal entropies, of the form $-P * \log P$. The product $-P * \log P$, which is Shannon's definition of marginal entropy, measures the amount of information gained when an event expected with probability P is observed. But here the equations for E and E′ express Shannon-like entropies, not true Shannon entropies, since EVE's entropies are signed (+ or −) and weighted (by H and R).

Actually, the equation for E is a *negative entropy* (total) for the set of possible events {win, loss} that occur at probabilities {P, Q}, since total entropy for a set of events is given by the sum of marginal entropies ($-P * \log P + -Q * \log Q$). When Eq. (1) is plotted (Fig. 4) this negative entropy is seen to be a bowl-shaped curve. E is minimal when the win and loss events are equally likely to occur (P=0.5) and E is maximal when one event is certain to occur (P=0 or P=1).

Interestingly, the equation for E′ is a mixture of oppositely signed and differently weighted marginal entropies, each of the form $H * P * \log P$ (and * R for the more general case where |R| <1). When Eq. (2) is plotted (Fig. 5) for

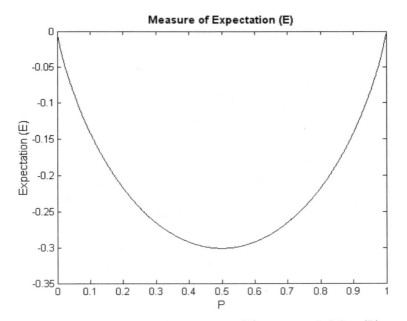

Fig. 4. EVE's measure of Expectation (E) versus probability (P)

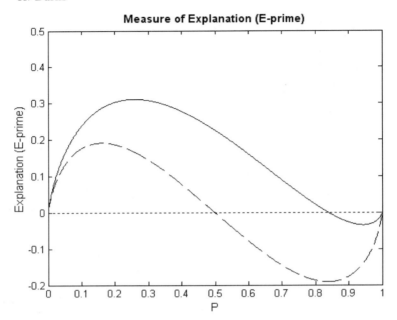

Fig. 5. EVE's measure of Explanation (E′) versus probability (P). Solid line is computed with $H^+/H^- = 2$. Dashed line is computed with $H^+/H^- = 1$

a typical ratio $H^+/H^- = 2$ (discussed below) then E′ is seen to be an S-shaped curve.

Finally, EVE′ sums E and E′ with weights called "Goldilocks" factors [1] denoted G and G′, which reflect the relative magnitude of pleasure or pleasure-prime obtained from a unit of E or E′, respectively. The resulting function for *aesthetic utility* (S) is written as follows and plotted in Fig. 6 for a typical ratio $G'/G = 3$ (discussed below):

$$S = G^*[P^* \log P + Q^* \log Q] - G'^*[H^{+*}P^* \log P - H^{-*}Q^* \log Q]. \quad (3)$$

As expected with G′>G, EVE's fun function (Fig. 6) is influenced by the bowl-shaped E (Fig. 4) but dominated by the S-shaped E′ (Fig. 5). Comparing Fig. 6 to Fig. 3 shows that the aesthetic utility (S) computed by EVE′ closely matches the apparent contribution of aesthetic utility to the economic utility of gambles in Prospect Theory.

Thus the S-shape assumed by a weighting function in Prospect Theory is actually derived as the pleasure function of an Affect Theory, EVE′. While this Affect Theory does assume values for *factors* (G and H), these values (constants) are less assumptive than Prospect Theory's *function* W and *factor* γ. As such, I believe that EVE′ offers new insights – through mathematical modeling of aesthetic utility, which in turn can be combined with economic utility to explain human choices in gambling and elsewhere.

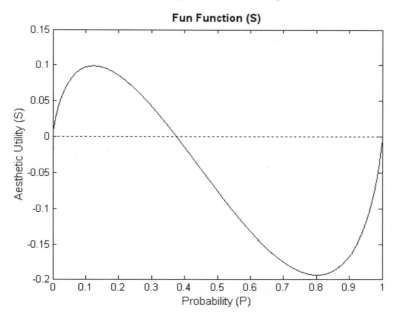

Fig. 6. EVE's aesthetic utility (S) versus probability (P), assuming $G'/G = 3$ and $H^+/H^- = 2$

Moreover, while G' and G (like H^+ and H^-) must be measured empirically, EVE' theoretically constrains the relation to be $G'>G$. This is because if $G'<G$ then the pleasure-prime gained at E' would always be less than the corresponding pleasure lost at E (leading to $V = -E$ and later E'). Thus $G'>G$ if the pleasure-prime gained at E' is ever to offset the loss at E to get net fun, as theorized in the E−E' tradeoff of EVE'. In this light the magnitude of G'/G (>1) can be seen as modeling a player's *call to adventure*, i.e., her willingness to give up some E for the rush of E'.

3.2 Economics and Energy

As derived above, EVE's fun function (S) is an equation for aesthetic utility *experienced* over time as wins and losses occur at frequencies of P and Q, respectively. Presumably players then internalize what is experienced so that S becomes an *expected* utility that governs their choices about whether to play or not. The derivation assumed the game was "fair slots", i.e., the expected economic utility was zero so aesthetic utility was the only force driving a person to play. This is the same as choosing between a sure-thing and an economically-equivalent gamble as discussed in Section 2 (see Figs. 1 and 2).

Now for gambles that are "not fair", choices will depend on the relative magnitudes of economic utility and aesthetic utility. For example, even if the game is fun a person will not play slots on an overpriced machine where

A $\gg$ P*J, since then the negative economic utility would outweigh the positive aesthetic utility of gambling. In other words, denoting the economic utility P*J − A by $ (which is zero for "fair slots"), there is some factor F by which aesthetic utility is weighed in summing it with economic utility to get a total *euphoric utility* U:

$$U = \$ + F^*S. \qquad (4)$$

This F is a weighting factor that must (like G and H) be obtained from empirical measurements. Also like G and H, F can be given a psychological interpretation where F is a person's *price for pleasure*, which determines how much a unit of aesthetic utility S is worth relative to a unit of economic utility $.

If F is high enough and the P of a gamble is such that S is positive (see Fig. 6), then the positive F*S can outweigh a negative $ so the total U is positive. Per EVE', it is this F*S that drives people to play even when the game is "not fair".

Of course not all people like to play slots, and this might be attributed to individual differences in F or G or H. These factors are theoretical limitations as well as practical advantages of EVE'. They are limitations because EVE' cannot predict human choices unless values for the factors (constants) are given or assumed. Yet the same factors are also advantages because they allow EVE' to model individual differences between people – within the fundamental regularities captured by the governing equations. These individual differences are not explicitly modeled by Prospect Theory or elsewhere in economics even though they are known to play a key role in human choices [16].

By explicitly modeling the factors F, G and H, as well as deriving a functional form for aesthetic utility S, EVE' can help to explicate what and how and why people buy and sell and trade – in real life as well as in games that simulate life [17].

Table 1 summarizes the *functions* of EVE', which model fundamental regularities (among people). Table 2 summarizes the *factors* of EVE', which model individual differences (between people).

Another advantage of EVE's theory is that aesthetic utility (S) can be given a physical interpretation, in which *enjoyment* comes from conversion

Table 1. Summary of EVE's *functions*, which model fundamental regularities (among people)

Function	Equation
Expectation	$E = \log P$
Violation	$V = -\log P$
Explanation	$E' = -\log P * H * R$
Aesthetic Utility	$S = G * E + G' * E'$
Economic Utility	$
Euphoric Utility	$U = \$ + F * S$

Table 2. Summary of EVE's *factors*, which model individual differences (between people)

Factor	Description
F	*price for pleasure*
G'/G	*call to adventure*
H⁺/H⁻	*sense of humor*

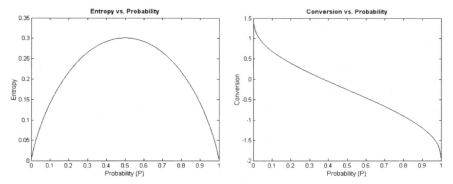

Fig. 7. (a) Left plot shows entropy function. (b) Right plot shows conversion function

of *entropy* to *energy*. To see this, note that EVE's summing of pleasure and pleasure-prime in S can be written as the product of two terms, one term representing entropy and the other term representing a conversion of entropy to energy in "sense-making" that leads to pleasure and pleasure-prime. Re-arranging Eq. (3), and using E = P * log P + Q * log Q, we obtain the following:

$$S = -[G^*E]^* \quad (5)$$
$$[(H^{+*}G'/G - 1)^*P^* \log P - (H^{-*}G'/G + 1)^*Q^* \log Q]/E,$$

where the first term $-[G^*E]$ is an *entropy* function and the second term is a *conversion* function. Further insight is gained when these two functions are plotted. Fig. 7a shows that the entropy function, which is a G-weighted entropy, exhibits a hump-shape. Fig. 7b shows that the conversion function (in the case of slots) is a monotonically decreasing function of P. When the two are combined to compute fun (S), as energy = entropy * conversion, we obtain the S-shape of Fig. 6 (and Fig. 3).

This S-shape is important because the same shape arises from the product of an entropy function (hump-shaped) and any monotonically decreasing conversion function. As discussed below under "Musical Flow" there appear to be other cases (perhaps many) where conversion is a monotonically decreasing function of P, hence the S-shape may be a universal characteristic of cognitive aesthetics in media experiences.

The physical interpretation of enjoyment (fun) as energy = entropy * conversion is useful because it may help to unify the study of aesthetics across various media, such as gaming [3], music [14] and artwork [15]. The same notion is practical because it may help to implement aesthetics in artificial agents that can calculate entropy and its conversion, hence compute energy of the sort that humans feel in fun.

The physical interpretation of "fun" as an energy resulting from entropy conversion can also be seen as relating to "flow" in what is sometimes called "optimal experience". Unlike EVE's theory of *fun*, the theory of *flow* [18] is concerned with personal not computational psychology. Nevertheless, flow is said to be optimal when the challenge (task) suits the person (skill), and this can be given a computational interpretation. According to EVE', the challenge presents some level of *entropy* that the person deals with in a *conversion* process (play or work) that yields *energy* – felt as fun or flow. This energy will be optimized when the product of entropy * conversion is maximized. In the case of slots, illustrated in Fig. 6, energy is optimized when the machine's payoff probability P is set at around 0.10–0.15.

As a benchmark of EVE's ability to predict fun or flow, it is useful to compare this predicted P for peak fun to the actual P at which payoffs are made by real casino slot machines. Presumably casinos have empirically optimized the Ps and Js by trial-and-error over the years to field machines that players find "most fun" – i.e., to maximize the hours played and hence the House profits. Note that the resulting win-win phenomenon, for aesthetics-economics of consumers-casinos, is theoretically possible and psychologically rational only when the analysis considers aesthetic utility as well as economic utility.

A prototypical example is the classic "twenty-one-bell three-wheel" machine for which detailed statistics are available. Scarne [19] notes that the House rake is 6%, and says (pp. 445-448): "I'd be playing it just for fun; I wouldn't expect to beat it in the long run." Like most slot machines there are N possible payoffs $\{J_1, J_2, ..., J_N\}$, each with a probability P of occurrence $\{P_1, P_2, ..., P_N\}$. In this case N=8 and the aggregate P computed from Scarne's statistics is equal to 0.13. Referring to Fig. 6 (and Fig. 3), this P=0.13 compares remarkably well to the range of P (0.10–0.15) in which EVE's function S is peaked.

Of course Fig. 6 applies only to players with a good *sense of humor* given by $H^+/H^- = 2$ and *call to adventure* $G'/G = 3$ as these individual preferences apply to slot play. But the same S-shape plotted in Fig. 6 has been measured in other gambling studies (see Fig. 3), which tends to support the assumed *sense of humor* ($H^+/H^- = 2$) as applying across people. Plus the same S-shape has been derived in music studies (see Fig. 8), which tends to support the assumed *call to adventure* ($G'/G = 3$) as a constant across domains. Therefore while EVE's curves are admittedly sensitive to individual differences modeled by G and H ratios, the apparent similarities (Figs. 3, 6 and 8) speak to EVE's generality as a computational paradigm for modeling aesthetics.

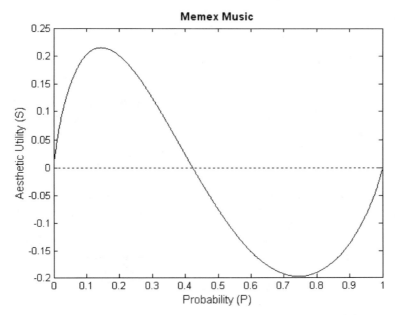

Fig. 8. EVE's measure of aesthetic utility (S) versus probability (P) for Memex music

3.3 Believing in Bayes

Recall that besides the factors F, G and H discussed above, EVE's fun function (S) also includes a resolution factor R. Here I wish to say more about R since it is essential to EVE's Bayesian entropies – and distinguishes EVE' from other information theories in computational aesthetics.

In the case of slots, R takes on a different sign (+ or −) for different outcomes (win or loss) depending on their valence, e.g., R = +1 for a win and R = −1 for a loss in Eq. (2) for E'. This gives rise to the S-shape of E', which in turn gives rise to the S-shape for fun when E' and E are combined via S = G*E + G'*E'.

In other cases R may affect not only the valence (+ or −) of an Explanation but also the magnitude of Explanation, i.e., when |R| <1. This is because R measures the extent to which an Explanation (E') resolves a Violation (V) of Expectation (E), and in EVE' [1] this resolution 0≤R≤1 is given by a Bayesian *posterior*. Henceforth ρ will be used to denote the posterior and other Bayesian beliefs (probabilities) in "mental models" [20] of human beings. These ρ (especially the ρ=R in E') are important because they distinguish EVE' from previous proposals that use information theory to model aesthetic experiences in artwork [21], music [22] and gaming [23].

Here it is useful to review the basics of Bayesian belief, starting with the difference between a *likelihood* and a *posterior* probability. The two are related by Bayes Rule via a *prior* probability as follows:

$$\rho(Y|e) = \rho(Y)^* \rho(e|Y)/\rho(e), \qquad (6)$$

where Y denotes a hypothesis (model) and e denotes some evidence (data). The equation says that a *posterior* probability $\rho(Y|e)$ of a hypothesis in light of evidence is given as the product [normalized by $\rho(e)$] of two terms: a *prior* probability $\rho(Y)$ of the hypothesis (before the evidence is obtained), and a *likelihood* $\rho(e|Y)$ that the observed evidence could have come from the hypothesized cause.

Note the operative word here is *cause*, as causal knowledge [24] is key to making inferences of the sort required to form Explanations (E'). The world works from cause to effect where causes (Y) are what lead to effects (e), so human knowledge about the world takes the form of likelihoods that express the probability $\rho(e|Y)$ of *effect given cause*. But in forming Explanations (E') the problem is exactly the reverse [13, 25], i.e., it is to infer a most likely *cause given effect* as expressed by a posterior probability $\rho(Y|e)$.

Numerically the posterior and likelihood are typically not equal since mathematically they differ by the prior in Bayes Rule, Eq. (6). This is extremely important, and failure to distinguish posteriors from likelihoods in probabilistic analyses is common but wrong [26]. In particular, previous applications of information theory to computational aesthetics have been limited because they have not adequately distinguished between causes (models) and effects (data) – sometimes treating data as both cause and effect. The details are discussed elsewhere [1, 15]. Here in the case of slots there are only two events (win or loss) and two causes ("good luck" or "bad luck"), simplifying the analysis of Bayesian belief and resolution (R) such that it is almost trivial, as follows.

First, using K to denote "good luck" and $\sim$K to denote "bad luck", the priors for the causes of events (win=ω or loss=λ) are given by $\rho(K)$=P and $\rho(\sim K)$=Q=1−P, where P is the probability of a payoff (win) from the slot machine. The likelihoods are similarly simplified, at least for a player (assumed here) who believes that good luck is the cause of a win and bad luck is the cause of a loss. That is, $\rho(\omega|K)$=1, $\rho(\omega|\sim K)$=0, $\rho(\lambda|\sim K)$=1, $\rho(\lambda|K)$=0. Substituting these priors and likelihoods into Bayes Rule yields the following posteriors: $\rho(K|\omega)$=1, $\rho(\sim K|\omega)$=0, $\rho(\sim K|\lambda)$=1, $\rho(K|\lambda)$=0. So for this case where likelihoods are either zero or one, each posterior is equal to the corresponding likelihood. Therefore the Explanation (resolution) for a win is "good luck", $\rho(K|\omega)$=1; likewise the Explanation (resolution) for a loss is "bad luck", $\rho(\sim K|\lambda)$=1.

To summarize, the Explanation (E') for each event (win or loss) in slots reflects a complete resolution (posterior) R of magnitude 1. This is why a term for R did not appear in the earlier expression for E', hence why R's role was limited to the valence (+ or −) of the Explanation. But as noted above, slots are a simplified and degenerate case, since in general $|R| \leq 1$ such that both the magnitude and valence of Explanation will be important to EVE's modeling (see "Musical Flow" below).

To reiterate, EVE' is a Bayesian-information theory of aesthetic experience where the Bayesian component (E') is critical. For slots the valence of R gives rise to the characteristic S-shape for fun, and in more complex cases the magnitude of R will also play a key role. This Bayesian component is what leads to the signed and weighted entropies by which EVE' models fun or flow, and these Bayesian entropies are similar to but not the same as Shannon entropies. That is why EVE' is characterized as a Bayesian-information theory, which goes further than previous information theories to advance a science of fun.

4 EVE's Extension

To generalize the contribution of EVE' beyond gambling, this section briefly reviews how the theory has been applied to music and artwork, and how it might be applied to other work in "serious" games.

4.1 Musical Flow

Music poses unique challenges to computational aesthetics since unlike other media (e.g., language and graphics) music does not typically represent objects or events outside the medium. Thus it is not clear how a listener might form an *Explanation* of a musical passage and indeed this has been the subject of much study. Nevertheless, Explanations (E') in *posterior* "realizations" [27] as well as Expectations (E) in *prior* "anticipations" [28] are clearly important to music perception.

In a similar vein, recent studies with EVE' found parallels between gambling and music in a composition called "Memex" [29]. This music was created by a computer operating on a database of classical compositions – selecting notes in a stochastic sequence governed by probabilities P and Q=1−P much like the sequence of events in slot play. Like slots, P affects the *entropy* of the composition (Memex) as well the *conversion* of this entropy to energy (in listeners). Unlike slots there is no valence (+ or −) from "win" or "loss" events as people listen, but like slots the magnitude of resolution (R) in E' plays a key role in shaping the form of EVE's fun function (S) for Memex. Again this highlights the importance of Bayesian belief in EVE'.

More details are presented elsewhere [14] but the main point is that R for Memex can be characterized as a monotonically decreasing function of P (although not identical to Fig. 7b). The conversion function then combines with a hump-shaped entropy function (Fig. 7a) to produce the S-shaped energy function shown in Fig. 8. The result is remarkably similar to Figs. 3 and 6, including the range of P where S is peaked. Of special note here is that the

P actually used by the Memex machine (composer) was adjusted by a human being (creator) until the music sounded "best" to him – which was at a P of 0.13! In short, the creator, who at the time did not know of EVE's model for fun in slots, settled on a P that is *exactly* the same as the P of a prototypical slot machine (see above) – and in the same range as EVE' predicts for peak fun in slots – and in the same range as EVE' predicts for peak S in Memex.

This finding further supports EVE's overall modeling of aesthetic utility. It also offers support from a different domain (music versus gambling) for the assumed *call to adventure*, since EVE's analysis of Memex used the same ratio $G'/G = 3$ that was assumed in the earlier analysis of slots.

4.2 Visual Art

Another application of EVE' has been to perception of visual images. This work has shown that EVE's computational theory reflects the philosophical theory of Aristotle's poetics [30]. In particular, Aristotle claimed that *imitation* is key to all art, where an aesthetic experience comes largely from *recognizing* the imitation after a set-up and surprise governed by *probability* and *causality*. The parallels to EVE' are obvious and applications of EVE' to graphics (along the lines of Aristotle's poetics) have illustrated some key points that were overlooked in previous applications of information theory to visual images [21, 31].

The details are discussed elsewhere in [1] and [15] but here it is useful to highlight one feature of EVE's entropies. That is, the expressions derived and discussed earlier (see Table 1) for E, V, E' and S are *average* quantities, where terms of the form log P are multiplied by P because P is the average rate at which an event occurs in a media experience over time – in a series of turns or a sequence of notes. In some cases, such as the visual images in [1] and [15], it is not an *average* but rather the *atomic* (instantaneous, not multiplied by P) measures of E−V−E' that must be computed to characterize the aesthetic experience.

In such cases, where the analysis is done at an atomic level rather than an average level, EVE's Bayesian entropies look even more different from the familiar Shannon entropies. Thus as noted earlier for the average cases of slot play and Memex, EVE's Bayesian entropies can be related – but not equated – to Shannon entropies of the sort proposed in earlier studies of computational aesthetics.

4.3 Serious Games

As a final extension of EVE' it is important to recognize that aesthetics are not limited to recreation. That is, although *fun* refers to *enjoyment* in play, the same aesthetic applies to any work where *flow* occurs in *engagement*.

This is especially true of "serious" games [17] where the purpose is to teach or train or try out new things (systems, techniques, etc.) in "hands on" simulations of real world situations – e.g., the exercises "played" in military, business, medical, government and other domains [32]. These games are often advertised as interactive and immersive, especially when they are played on computers. But to be effective they must be more than that; they must be engaging [33] and even fun [11], especially if players are to "learn" what the game is meant to teach.

Today's designers of serious games face much the same problem as designers of commercial games, namely they are lacking a science of flow (engagement) – which is also the science of fun (enjoyment). As argued earlier, the underlying aesthetic is essentially the same so the fundamental science is also the same.

The application of this science also extends beyond just games, serious or otherwise, to all work. This is because today's workers are often employed in computerized tasks and yet they are not always so engaged in this work [34]. Said another way, their flow [18] is often far less than an "optimal experience" for them as well as their employers.

Previous research aimed at improving engagement between people and computers has focused mostly on *usability* from an ergonomic perspective – rather than *likeability* from an aesthetic perspective. One reason may be that only recently has the importance of *aesthetics* to ergonomics and economics become apparent. Another reason may be that only recently has the need for *computing* aesthetics been recognized as a vital aspect of designing computers that can effectively engage human beings as they play along (and against) each other in work as well as games.

But now the time has come for a science of fun.

5 Conclusion

As argued at the start, intelligence and aesthetics cannot be separated since each affects the other in cognition-emotion interactions. EVE's contribution is a new paradigm for implementing computational aesthetics in artificial intelligence.

EVE$'$ does so with a *Bayesian-information* approach that goes beyond earlier *Shannon-information* models to better address the fundamental philosophy of Aristotelian aesthetics. EVE$'$ also goes beyond the *economic utility* of previous theories like Prospect Theory in mathematical modeling of *aesthetic utility*. The result is a formal gauge of fun in games, which can extend to other arts and any work.

Thus EVE$'$ is an important start – towards a science of fun.

References

1. Burns, K.: Atoms of EVE': A Bayesian Basis for Aesthetic Analysis of Style in Sketching. Artificial Intelligence for Engineering Design, Analysis and Manufacturing, Vol. 20 (2006) 185–199
2. Huhtamo, E.: Slots of Fun, Slots of Trouble: An Archaeology of Arcade Gaming. In: Raessens, J., Goldstein, J. (eds.): Handbook of Computer Game Studies. MIT Press, Cambridge MA (2005)
3. Burns, K.: Fun in Slots. Proceedings of the IEEE Conference on Computational Intelligence in Games, Reno NV (2006)
4. De Martino, B., Kumaran, D., Seymour, B., Dolan, R.: Frames, Biases and Rational Decision-making in the Human Brain. Science, Vol. 313 (2006) 684–687
5. Tversky, A., Kahneman, D.: Advances in Prospect Theory: Cumulative Representation of Uncertainty. Journal of Risk and Uncertainty, Vol. 5 (1992) 297–323. In: Kahneman, D., Tversky, A. (eds.): Choices, Values and Frames. Cambridge University Press, Cambridge UK (2000) 44–65
6. Prelec, D.: Compound Invariant Weighting Functions in Prospect Theory. In: Kahneman, D., Tversky, A. (eds.): Choices, Values and Frames. Cambridge University Press, Cambridge UK (2000) 67–92
7. Turner, M. (ed.): The Artful Mind: Cognitive Science and the Riddle of Human Creativity. Oxford University Press, Oxford UK (2006)
8. Postrel, V.: The Substance of Style: How the Rise of Aesthetic Value is Remaking Commerce, Culture and Consciousness. Harper Collins, New York NY (2003)
9. Liu, H., Mihalcea, R. (eds.): Computational Aesthetics: Artificial Intelligence Approaches to Beauty and Happiness. Technical Report WS-06-04, AAAI Press, Menlo Park CA (2006)
10. Salen, K., Zimmerman, E.: Rules of Play: Game Design Fundamentals. MIT Press, Cambridge MA (2004)
11. Koster, R.: A Theory of Fun for Game Design. Paraglyph Press, Scottsdale AZ (2005)
12. Shannon, C., Weaver, W.: The Mathematical Theory of Communication. University of Illinios Press, Urbana IL (1949)
13. Knill, D., Richards, W. (eds.): Perception as Bayesian Inference. Cambridge University Press, Cambridge UK (1996)
14. Burns, K., Dubnov, S.: Memex Music and Gambling Games: EVE's Take on Lucky Number 13. Proceedings of the AAAI Workshop on Computational Aesthetics, Technical Report WS-06-04, AAAI Press, Menlo Park CA (2006) 30–36
15. Burns, K.: Bayesian Beauty: On the Art of EVE' and the Act of Enjoyment. Proceedings of the AAAI Workshop on Computational Aesthetics, Technical Report WS-06-04, AAAI Press, Menlo Park CA (2006) 74–78
16. Frederick, S.: Cognitive Reflection and Decision Making. Journal of Economic Perspectives, Vol. 19, No. 4 (2005) 25–42
17. Abt, C.: Serious Games. Viking Press, New York NY (1970)
18. Csikszentmihalyi, M.: Flow: The Psychology of Optimal Experience. Harper Collins, New York NY (1991)
19. Scarne, J.: Scarne's New Complete Guide to Gambling. Simon & Schuster, New York NY (1961)
20. Burns, K.: Mental Models of Line Drawings. Perception, Vol. 30, No. 6. (2001) 1249–1261

21. Bense, M.: Aesthetica: Einführung in die Neue Aesthetik. Agis-Verlag, Baden-Baden (1965)
22. Dubnov, S., McAdams, S., Reynolds, R.: Predicting Human Reactions to Music on the Basis of Similarity Structure and Information Theoretic Measures of Sound Signal. Proceedings of the AAAI Symposium on Style and Meaning in Language, Music, Art and Design, Technical Report FS-04-07, AAAI Press, Menlo Park CA (2004) 37–40
23. Yannakakis, G., Lund, H., Hallam, J.: Modeling Children's Entertainment in the Playwear Playground. Proceedings of the IEEE Conference on Computational Intelligence in Games, Reno NV (2006)
24. Pearl, J.: Causality: Models, Reasoning and Inference. Cambridge University Press, Cambridge UK (2000)
25. Burns, K.: Mental Models and Normal Errors. In: Montgomery, H., Lipshitz, R., Brehmer, B. (eds.): How Professional Make Decisions. Lawrence Erlbaum, Mahwah NJ (2005)
26. Burns, K.: Bayesian Inference in Disputed Authorship: A Case Study of Cognitive Errors and a New System for Decision Support. Information Sciences, Vol. 176 (2006) 1570–1589
27. Narmour, E.: The Analysis and Cognition of Basic Melodic Structures: The Implication-Realization Model. University of Chicago Press, Chicago IL (1990)
28. Huron, D.: Sweet Anticipation: Music and the Psychology of Expectation. MIT Press, Cambridge MA (2006)
29. Dubnov, S.: Thoughts About Memex. http://music.ucsd.edu/~sdubnov
30. Butcher, S.: Aristotle Poetics. Dover, New York NY (1951)
31. Feldman, J., Singh, M.: Information Along Contours and Object Boundaries. Psychological Review, Vol. 112, No. 1 (2005) 243–252
32. Aldrich, C.: Learning by Doing: A Comprehensive Guide to Simulations, Computer Games and Pedagogy in e-Learning and Other Educational Experiences. John Wiley & Sons (2005)
33. Prensky, M.: Fun, Play and Games: What Makes Games Engaging. In Prensky, M: Digital Game-Based Learning, Chapter 5. McGraw Hill (2001)
34. Hoffman, R., Hayes, P.: The Pleasure Principle. IEEE Intelligent Systems, January/February (2004) 86–89

Capturing Player Enjoyment in Computer Games

Georgios N. Yannakakis and John Hallam

Maersk Mc-Kinney Moller Institute
University of Southern Denmark
Campusvej 55, Odense M, DK-5230
{georgios,john}@mip.sdu.dk

The current state-of-the-art in intelligent game design using Artificial Intelligence (AI) techniques is mainly focused on generating human-like and intelligent characters. Even though complex opponent behaviors emerge through various machine learning techniques, there is generally no further analysis of whether these behaviors contribute to the satisfaction of the player. The implicit hypothesis motivating this research is that intelligent opponent behaviors enable the player to gain more satisfaction from the game. This hypothesis may well be true; however, since no notion of entertainment or enjoyment is explicitly defined, there is therefore no evidence that a specific opponent behavior generates enjoyable games.

This chapter introduces a discussion of quantitative entertainment capture in real-time and presents two dissimilar approaches for modeling player satisfaction. Successfully capturing the level of entertainment during play provides insights for designing the appropriate AI methodology for entertainment augmentation in real-time. For this purpose, adaptive on-line learning methodologies are proposed and their strengths and limitations are discussed.

1 Introduction

Cognitive modeling within human-computer interactive systems is a prominent area of research. Computer games, as examples of such systems, provide an ideal environment for research in artificial intelligence (AI), because they are based on simulations of highly complex and dynamic multi-agent worlds [1, 2, 3]. Moreover, computer games offer a promising ground for cognitive modeling since they embed rich forms of interactivity between humans and non-player characters (NPCs) [4]. Being able to capture quantitatively the level of user (gamer) engagement or satisfaction in real-time can grant insights to the appropriate AI methodology for enhancing the quality of playing experience [5] and furthermore be used to adjust digital entertainment environments according to individual user preferences.

Motivated by the lack of quantitative models of entertainment, endeavors to measure and augment player satisfaction in real-time are presented in this chapter. More specifically, the open question of modeling entertainment during game play is discussed and the strengths and weaknesses of previous attempts in the field are outlined. Herein entertainment is defined qualitatively primarily as the level of satisfaction generated by the real-time player-game opponent interaction — by 'opponent' we define any controllable interactive feature of the game. We view a game primarily as a learning process, and the level of entertainment is kept high when game opponents enable new learning patterns ('not too easy a game') for the player that can be perceived and learned by the player ('not too difficult a game') [6, 7]. On the same basis, according to [8] — within the axis of emotions varying from boredom to fascination — learning is highly correlated with interest, curiosity and intrigue perceived. The collection of these emotions is referred to as entertainment (or "fun") in this chapter.

Two different approaches for quantitatively capturing and enhancing the real-time entertainment value of computer games are presented in this chapter: one based on empirical design of entertainment metrics and one where quantitative entertainment estimates are extracted using machine learning techniques, grounded in psychological studies. The predator/prey game genre is used for the experiments presented here; though it is argued that the proposed techniques can be applied more generally, to other game genres.

In the first, "empirical," approach, a quantitative metric of the 'interestingness' of opponent behaviors is designed based on qualitative considerations of what is enjoyable in predator/prey games. A mathematical formulation of those considerations, based upon data observable during game play, is derived. This metric is validated successfully against the human notion of entertainment.

In the second approach, entertainment modeling is pursued by following the theoretical principles of Malone's [9] intrinsic qualitative factors for engaging game play, namely *challenge* (i.e. *'provide a goal whose attainment is uncertain'*), *curiosity* (i.e. *'what will happen next in the game?'*) and *fantasy* (i.e. *'show or evoke images of physical objects or social situations not actually present'*) and driven by the basic concepts of the Theory of Flow [10] (*'flow is the mental state in which players are so involved in the game that nothing else matters'*). Quantitative measures for challenge and curiosity are inspired by the "empirical" approach to entertainment metrics. They are represented by measures computed from appropriate game features based on the interaction of player and opponent behavior. A mapping between the quantitative values of these challenge and curiosity measures and the human notion of entertainment is then constructed using evolving neural networks (NNs).

The chapter concludes with a discussion of several remaining open questions regarding entertainment modeling and proposes future directions to answer these questions. The limitations of the presented methodology, and the extensibility of the proposed approaches of entertainment capture and augmentation to other genres of digital entertainment, are also discussed.

2 Capturing Entertainment

There have been several psychological studies to identify what is "fun" in a game and what engages people playing computer games. Theoretical approaches include Malone's principles of intrinsic qualitative factors for engaging game play [9], namely challenge, curiosity and fantasy as well as the well-known concepts of the theory of flow [10] incorporated in computer games as a model for evaluating player enjoyment, namely *GameFlow* [11].

A comprehensive review of the literature on qualitative approaches for modeling player enjoyment demonstrates a tendency to overlap with Malone's and Csikszentmihalyi's foundational concepts. Many of these approaches are based on Lazzaro's "fun" clustering which uses four entertainment factors based on facial expressions and data obtained from game surveys of players [12]. According to Lazzaro, the four components of entertainment are: hard fun (related to the challenge factor of Malone), easy fun (related to the curiosity factor of Malone), altered states (i.e. 'the way in which perception, behavior, and thought combine in a collective context to produce emotions and other internal sensations' — closely related to Malone's fantasy factor) — and socialization (the people factor). Koster's [7] theory of fun, which is primarily inspired by Lazzaro's four factors, defines "fun" as the act of mastering the game mentally. An alternative approach to fun measure is presented in [13] where fun is composed of three dimensions: endurability, engagement and expectations. Questionnaire tools and methodologies are proposed in order to empirically capture the level of fun for evaluating the usability of novel interfaces with children.

Kline and Arlidge [14] support Lazzaro's findings since their studies on on-line massive multi-player games (e.g. Counter-Strike) identify the socialization factor as an additional component to Malone's factors for "fun" game play experiences. Their clustering of player styles ('player archetypes') corresponds to these four dimensions of entertainment: *Warriors* are those who prioritize combat features and realism (closely related to Malone's fantasy factor), *Narrators* are those who place priority on the plot, characters, and exploration but they do not like games that are challenging (closely related to Malone's curiosity factor), *Strategists* are those that prioritize complex strategies, challenging game play and mastery (closely related to Malone's challenge factor) and *Interactors* for whom competition and cooperation with other players is of the highest importance (socialization factor).

On that basis, some endeavors towards the criteria that collectively make simple on-line games appealing are discussed in [15]. The human survey-based outcome of that work presents challenge, diversity and unpredictability as primary criteria for enjoyable opponent behaviors.

Vorderer et al. [4] present a quantitative analysis (through an extensive human player survey on the Tomb Raider game) of the impact of competition (i.e. challenge) on entertainment and identify challenge as the most important determinant of the enjoyment perceived by video game players. They claim that a successful completion of a task generates sympathetic arousal, especially when the challenge of the task matches the player's abilities. In their analysis, social competition (just like Lazzaro's people factor and Kline's and Arlidge's *Interactors*) appears to enhance entertainment. Their survey, however, is based on single game-task enjoyment evaluations rather than comparative evaluations of several scenarios.

The study by Choi et al. [16] ranks perceptual and cognitive fun as the top-level factors for designing "fun" computer games. For the former, it is the game interface that affects the player's perception (vividness and imagination) — this corresponds to Malone's fantasy factor. For the latter, it is the game mechanics (level of interactivity) that affect the player's cognitive process — which comprises the challenge and satisfaction factors. According to Choi et al. (*ibid.*), challenge and satisfaction appear as independent processes, in contrast to the views of Malone [9] and Yannakakis et al. [17] where satisfaction derives from the appropriate level of challenge and other game components. Moreover, in Choi et al.'s study, vividness and imagination (perceptual fun) appear more important entertainment factors for players of strategic simulation games and challenge and satisfaction (cognitive fun) appear more important for role-playing games.

As previously mentioned, research in the field of game AI is mainly focused on generating human-like (believable) and intelligent (see [3, 18] among others) characters. Complex NPC behaviors can emerge through various AI techniques; however, there is no further analysis of whether these behaviors have a positive impact to the satisfaction of the player during play. According to Taatgen et al. [19], believability of computer game opponents, which are generated through cognitive models, is strongly correlated with enjoyable games. Such implicit research hypotheses may well be true; however, there is little evidence that specific NPCs generate enjoyable games unless a notion of interest or enjoyment is explicitly defined.

Iida's work on metrics of entertainment in board games was the first attempt in the area of quantitative "fun" modeling. He introduced a general metric of entertainment for variants of chess games depending on average game length and possible moves [20]. Other work in the field of quantitative entertainment capture is based on the hypothesis that the player-opponent interaction — rather than the audiovisual features, the context or the genre of the game — is the property that contributes the majority of the quality features of entertainment in a computer game [6]. Based on this fundamental assumption, a metric for measuring the real time entertainment value of predator/prey games was designed, and established as efficient and reliable by validation against human judgement [21, 22]. Further studies by Yannakakis and Hallam [23] have shown that Artificial Neural Networks (ANN) and fuzzy neural networks can extract a better estimator of player satisfaction than a custom-designed (or designer-driven) one, given appropriate estimators of the challenge and curiosity of the game and data on human players' preferences. Similar work in adjusting a game's difficulty include endeavors through reinforcement learning [24], genetic algorithms [25], probabilistic models [26] and dynamic scripting [27]. However, the aforementioned attempts are based on the assumption that challenge is the only factor that contributes to enjoyable gaming experiences while results reported have not been cross-verified by human players.

A step further to entertainment capture is towards games of richer human-computer interaction and affect recognizers which are able to identify correlations between physiological signals and the human notion of entertainment. Experiments by Yannakakis et al. [28] have already shown a significant correlation of average heart rate with children's perceived entertainment in action games played in interactive physical playgrounds. Moreover, Rani et al. [29] propose a methodology for detecting anxiety level of the player and appropriately adjusting the level of challenge in the game of 'Pong' based on recorded physiological signals in real-time and subject's self-reports of their emotional experiences during game play. Physiological state (heart-rate, galvanic skin response) prediction models have also been proposed for potential entertainment augmentation in computer games [30].

Following the theoretical principles reported from Malone [9], Koster [7] and Yannakakis [21], and to a lesser degree from Lazzaro [12] and Kline and Arlidge [14], this chapter is primarily focused on the contributions of game opponents' behavior (by enabling appropriate learning patterns on which the player may train [7]) to the real-time entertainment value of the game. This chapter therefore excludes the socialization factor of entertaining game play and investigates instead entertainment perceived in single player scenarios. We argue that among the three dimensions of "fun" (endurability, engagement, expectations) defined in [13] it is only *engagement* that is affected by the opponent since both *endurability* and *expectations* are based primarily on the game design *per se*. Given a successful interactive game design that yields high expectations and endurability, we only focus on the level of engagement that generates fun (entertainment).

Rather than being based purely on theoretical assumptions and visual observations of players' satisfaction, this chapter presents two different approaches that attempt to capture quantitatively the level of player entertainment in computer games. First, a custom-designed quantitative metric of entertainment is proposed, motivated by qualitative considerations of what is enjoyable in computer games. The metric has been validated against humans' preferences. Second, a mapping between estimators of Malone's challenge and curiosity entertainment factors and humans' valuations of entertainment is derived using evolving NNs.

3 The Test-bed Game

The test-bed studied here is a modified version of the original Pac-Man computer game released by Namco. The player's (*PacMan's*) goal is to eat all the pellets appearing in a maze-shaped stage while avoiding being killed by the four *Ghosts*. The game is over when either all pellets in the stage have been eaten by *PacMan*, *Ghosts* manage to kill *PacMan* or a predetermined number of simulation steps is reached without either of the above occurring. In the last case, the game restarts from the same initial positions for all five characters. In the test-bed game, *PacMan* is controlled by the human player while a multi-layered feedforward neural controller is employed to manage the *Ghosts'* motion.

The game is investigated from the opponents' viewpoint and more specifically how the *Ghosts'* adaptive behaviors and the levels of challenge and curiosity they generate can collectively contribute to player satisfaction. The game field (i.e. stage) consists of corridors and walls. Both dimensions and maze structure of the stage are predefined. For the experiments presented in this chapter we use a 19×29 grid maze-stage with corridors 1 grid-cell wide (see [31] for more details of the Pac-Man game design).

We choose predator/prey games as the initial genre for this research since, given our aims, they provide us with unique properties: in such games we can deliberately abstract the environment and concentrate on the characters' behavior. Moreover, we are able to easily control a learning process through on-line interaction. Other genres of game (e.g. first person shooters) offer similar properties; however predator/prey games are chosen for their simplicity as far as their development and design are concerned.

4 Empirical Estimation of Entertainment

As noted in section 3, predator/prey games will be our test-bed genre for the investigation of enjoyable games. More specifically, in the games studied, the prey is controlled by the player and the predators are the computer-controlled opponents (non-player characters, or NPCs).

In the approach presented in this section, a quantitative metric of player satisfaction is designed based on general criteria of enjoyment. The first step towards generating enjoyable computer games is therefore to identify the criteria or features of games that collectively produce enjoyment (or else interest) in such games. Second, quantitative estimators for these criteria are defined and combined, in a suitable mathematical formula, to give a single quantity correlated with player satisfaction (interest). Finally, this formulation of player interest is tested against human players' judgement in real conditions using the Pac-Man test-bed (see section 4.2).

Following the principles of Yannakakis and Hallam [6, 22] we will ignore mechanics, audiovisual representation, control and interface contributions to the enjoyment of the player and we will concentrate on the opponents' behaviors. A well-designed and popular game such as Pac-Man can fulfil all aspects of player satisfaction incorporated in the above-mentioned design game features. The player, however, may contribute to his/her own entertainment through interaction with the opponents of the game and therefore is included implicitly in the interest formulation presented here — see also [32] for studies of the player's impact on his/her entertainment.

Criteria. By observing the opponents' behavior in various predator/prey games we attempted to identify the key features that generate entertainment for the player. These features were experimentally validated against various opponents with different strategies and redefined when appropriate. Hence, by being as objective and generic as possible, we believe that the criteria that collectively define interest on any predator/prey game are as follows.

1. *When the game is neither too hard nor too easy.* In other words, the game is interesting when predators (opponents) manage to kill the prey (player) sometimes but not always. In that sense, given a specific game structure and a player, highly effective opponent behaviors are not interesting behaviors and *vice versa*.
2. *When there is diversity in opponents' behavior between games.* That is, the game is interesting when NPCs are able to find dissimilar ways of hunting and killing the player in each game so that their strategy is more variable.
3. *When opponents' behavior is aggressive rather than static.* That is, the game is interesting when the predators move constantly all over the game world and cover it uniformly. This behavior gives the player the impression of an intelligent strategy for the opponents.

Metrics. These three general criteria must now be expressed in quantitative terms using data observable during game play. We therefore let a group of game opponents — the number of opponents depends on the specific game under examination — play the game N times (each game for a sufficiently large evaluation period of t_{max} steps) and record the steps t_k taken to kill the player in each game k as well as the total number of visits v_{ik} opponents make to each cell i of the game grid.

Given these data, quantifications of the three interest criteria proposed above can be presented as follows.

1. **Appropriate Level of Challenge.** The game is uninteresting when either the opponents consistently kill the player quickly (game too hard) or when the game consistently runs for long periods (game too easy). This criterion can be quantified by T in (1.1) below.

$$T = [1 - (E\{t_k\}/\max\{t_k\})]^{p_1} \qquad (1.1)$$

where $E\{t_k\}$ is the average number of simulation steps taken to kill the prey-player over the N games; $\max\{t_k\}$ is the maximum t_k over the N games — $\max\{t_k\} \leq t_{max}$; and p_1 is a weighting parameter. T is high when $E\{t_k\}$ is low compared to $\max\{t_k\}$, that is, when games occur that are much longer than average.

p_1 is adjusted so as to control the impact of the bracketed term in the formula for T. By selecting values of $p_1 < 1$ we reward quite challenging opponents more than near-optimal killers, since we compress the T value toward 1. p_1 is chosen as 0.5 for the experiments presented here.

The T estimate of interest demonstrates that the greater the difference between the average and the maximum number of steps taken to kill the player, the higher the interest of the game. Given (1.1), both easy-killing ('*too easy*') and near-optimal ('*too hard*') behaviors receive low interest estimate values (i.e. $E\{t_k\} \simeq \max\{t_k\}$). This metric is also called 'challenge'.

2. **Behavior Diversity.** The game is interesting when the NPCs exhibit a diversity of behavior between games. One manifestation of this is that the time taken to kill the player varies between games. Thus a quantitative metric for this second criterion is given by S in (1.2) below.

$$S = (\sigma_{t_k}/\sigma_{max})^{p_2} \qquad (1.2)$$

where

$$\sigma_{max} = \frac{1}{2}\sqrt{\frac{N}{(N-1)}}(t_{max} - t_{min}) \qquad (1.3)$$

and σ_{t_k} is the standard deviation of t_k over the N games; σ_{max} is an estimate, based on the range of $\{t_k\}$ of the maximum value of σ_{t_k}; t_{min} is the minimum number of steps required for predators to kill the prey when playing against some 'well-behaved' fixed strategy near-optimal predators ($t_{min} \leq t_k$); and p_2 is a weighting parameter which is set so as σ_{t_k} has a linear effect on S ($p_2 = 1$).

The S increases proportionally with the standard deviation of the steps taken to kill the player over N games. Therefore, using S as defined here, we promote predators that produce high diversity in the time taken to kill the prey.

3. **Spatial Diversity.** The game is more interesting when opponents appear to move around actively rather than remain static or passively follow the player. A good measure for quantifying this criterion is through entropy of the opponents' visits to the cells of the game grid during a game, since the entropy quantifies the completeness and uniformity with which the opponents cover the stage. Hence, for each game, the cell visit entropy is calculated and normalized into $[0, 1]$ via (1.4).

$$H_n = \left[-\frac{1}{\log V_n}\sum_i \frac{v_{in}}{V_n}\log\left(\frac{v_{in}}{V_n}\right)\right]^{p_3} \qquad (1.4)$$

where V_n is the total number of visits to all visited cells (i.e. $V_n = \sum_i v_{in}$) and p_3 is a weighting parameter. p_3 is adjusted in order to promote very high H_n values and furthermore to emphasize the distinction between high and low normalized entropy values. Appropriate p_3 parameter values which serve this purpose are those greater than one ($p_3 = 4$ in this chapter), since they stretch the value of H_n away from 1.

Given the normalized entropy values H_n for all N games, the interest estimate for the third criterion can be represented by their average value $E\{H_n\}$ over the N games. This implies that the higher the average entropy value, the more interesting the game is.

The three individual criterion metrics defined above are combined linearly to produce a single metric of interest (equation 1.5) whose properties match the qualitative considerations developed above.

$$I = \frac{\gamma T + \delta S + \epsilon E\{H_n\}}{\gamma + \delta + \epsilon} \qquad (1.5)$$

where I is the interest value of the predator/prey game; γ, δ and ϵ are criterion weight parameters.

The approach to entertainment modeling represented by equation (1.5) is both innovative and efficient. However, it should be clear from the foregoing discussion that there are many possible formulae such as equation (1.5) which would be consistent with the qualitative criteria proposed for predator/prey games. Other successful quantitative metrics for the appropriate level of challenge, the opponents' diversity and the opponents' spatial diversity may be designed and more qualitative criteria may be inserted in the interest formula. Alternative mathematical functions for combining and weighting the various criteria could be employed.

For example, other metrics for measuring the appropriate level of challenge could be used: one could come up with a T metric assuming that the appropriate level of challenge follows a Gaussian distribution over $E\{t_k\}$ and that the interest value of a given game varies depending on how long it is — *very short* ($E\{t_k\} \approx t_{min}$) games tend to be frustrating and *long games* ($E\{t_k\} \approx max\{t_k\}$) tend to be boring. (However, very short games are not frequent in the experiments presented here and, therefore, by varying the weight parameter p_1 in the proposed T metric (see (1.1)) we are able to obtain an adequate level of variation in measured challenge.)

To obtain values for the interest formula weighting parameters γ, δ and ϵ we select empirical values based on the specific game in question. For Pac-Man, spatial diversity of the opponents is of the greatest interest: the game no longer engages the player when *Ghosts* stick in a corner instead of wandering around the stage. Thus, diversity in game play (S) and challenge (T) should come next in the importance list of interest criteria. Given the above-mentioned statements and by adjusting these three parameters so that the interest value escalates as the opponent behavior changes from randomly generated (too easy) to near-optimal hunting (too difficult) and then to following *Ghost* behaviors, we come up with $\gamma = 1, \delta = 2$ and $\epsilon = 3$.

The question remains, however, whether the number produced by such a formula really captures anything useful concerning a notion so potentially complex as human enjoyment. That question is addressed in section 4.2 below.

4.1 Generality of the Metric

The interest metric introduced in equation (1.5) can be applied in principle to any predator/prey computer game because it is based on generic measurable features of this category of games. These features include the time required to kill the prey and the predators' entropy in the game field. Thus, it appears that (1.5) — or a similar measure based on the same concepts — constitutes a generic interest approximation of predator/prey computer games. Evidence demonstrating the interest metric's generality appears in [33] through successful application of the I value metric to two quite dissimilar predator/prey games.

Moreover, given the two first interest criteria previously defined, the approach can be generalized to all computer games. Indeed, no player likes any computer game that is too hard or too easy to play and, furthermore, any player would enjoy diversity throughout the play of any game. The third interest criterion is applicable to games where spatial diversity is important which, apart from predator/prey games, may also include action, strategy and team sports games according to the computer game genre classification of Laird and van Lent [3]. As long as game designers can determine and extract the measurable features of the opponent behavior that generate excitement for the player, and identify observable indices of them that can be computed from data collected during game play, a mathematical formula can be designed in order to collectively represent them.

Finally, a validation experiment like that presented in section 4.2 below can be used to assess the performance of the designed formula or test variants of it.

4.2 Experiments

Given that the interest measure defined above has been constructed to reflect the designers' intuitions about those features of games that contribute to player interest and satisfaction, one would expect that games with higher values of I would be judged more satisfying by human players.

To investigate this, the Pac-Man game was used to acquire data on human judgement of entertainment. Thirty subjects (43.3% females, 56.7% males) whose age covered a range between 17 and 51 years participated in this experiment. In addition, all subjects spoke English (language of the survey questionnaire) as a foreign language since their nationality was either Danish (90%) or Greek (10%) [21, 22].

Subjects in the experiment played against five selected opponents differing in the I value they generate against a well-behaved computer-programmed player. Each player played several paired sets of games such that all pairwise combinations of the 5 opponents, in each order, were covered by the group of subjects (see [6] for details of the experimental design). For each pair, the player indicated which opponent generated the more satisfying game (players were not aware of the computed I values of the opponents they faced).

The degree of correlation between human judgement of entertainment and the computed I value is found by matching the entertainment preferences between the five opponents recorded by the human players and the I value ranking. According to the subjects' answers the I value is correlated highly with human judgement ($r = 0.4444$, p-value $= 1.17 \cdot 10^{-8}$ — see [21, 22]). These five opponents are used as a baseline for validating all entertainment modeling approaches presented in this chapter (see section 5.3).

In addition, players completed survey questions designed to elicit information about their likeliness for Pac-Man as a computer game, which allowed them to be grouped into three types: the ones that conceptually did not particularly like Pac-Man, the ones believed that Pac-Man is an interesting game and the ones that liked Pac-Man very much. It was found that neither the type of player nor the order in which opponents were encountered had any significant effect on the player's judgement of the relative interest of the games in a pair (see [21, 22] for full details of the analysis).

4.3 Conclusions

Given observations, intuitions and informal empirical studies on the predator/prey genre of games, generic criteria that contribute to the satisfaction for the player and map to measurable characteristics of the opponent behavior were defined.

According to the hypothesis of Yannakakis and Hallam [6], the player-opponent interaction — rather than the audiovisual features, the context, the mechanics, the control, the interface or the genre of the game — is the property that contributes the majority of the quality features of entertainment in a computer game. Given this fundamental assumption, a metric for measuring the real-time entertainment value of predator/prey games was motivated and designed. This value is built upon three generic entertainment criteria: appropriate level of challenge, opponents' behavior diversity and opponents' spatial diversity.

By testing predator/prey games with human players, it was confirmed that the interest value computed by equation (1.5) is consistent with human judgement of entertainment. In fact, the human player's notion of interest in the Pac-Man game seems to correlate highly with the computed interest metric, independently of player type or order of play [22].

Thus, we demonstrate the first approach to quantitative modeling of entertainment: the idea of using a custom-designed (but nevertheless quite generic) mathematical expression that yields a numerical value well-correlated with human judgement. The metric formula rests on measurable features of opponent behavior computable from data collected during game play.

5 Quantitative Analysis of Entertainment Factors Derived from Psychological studies

The second approach to entertainment capture is in a sense inverse to the empirical, designer-driven, estimation of entertainment presented in section 4. In that approach, generic criteria describing interesting games were identified by intuition and consideration of game experiences on the part of the entertainment metric designer.

Although shown to yield a result that correlates well with human judgement, a more satisfying approach might be to start with the theoretical work on qualitative factors of entertainment reviewed in section 2 — specifically, Malone's qualitative factors of entertainment: challenge and curiosity [9] — and attempt to construct a quantitative measure using them.

Quantitative measures for challenge and curiosity are inspired by previous work on entertainment metrics [6] and computed from data gathered during the real-time player-opponent interaction. The measured challenge and curiosity values are combined with human judgements on the relative entertainment of games using machine learning techniques to construct an expression analogous to equation 1.5 which is highly correlated with human choices. Again, the Pac-Man game is employed as a test-bed for the approach.

Two NN types, namely a feedforward NN and a fuzzy-neural network (fuzzy-NN), are used as alternatives to represent the entertainment metric. In each case, the inputs to the network are the measures of challenge and curiosity for a particular game (opponent). The output is an analogue of the I metric. The networks are thus used as function approximators for the expression defining the interest metric. Training of the approximators is done using artificial evolution.

The procedures used are described in more detail below. A comparison between the two alternatives is presented and the results are validated against and compared with the custom-designed metric of equation (1.5). The results of the study demonstrate that both NNs represent functions — possible interest metrics based on challenge and curiosity measures — whose qualitative features are consistent with Malone's corresponding entertainment factors. Furthermore, the evolved feedforward NN provides a more accurate model of player satisfaction for Pac-Man than the custom-designed model (the I value), presented in the previous section, for this genre of games [33].

5.1 Experimental Data

As a final part of the experiment presented in section 4.2, subjects played a set of 25 games against each of two well-behaved opponents (A and B). After the 50 games, the player records whether the first set or the second set of games was the more interesting, i.e. whether A or B generated a more interesting game. (The preference of A over B is also written as $A \succ B$.) To minimize order effects, each subject plays the aforementioned sets both orders during the experiment.

Given the recorded values of human playing times t_k over the 50 (2×25) games against a specific opponent, either A or B, the average playing time ($E\{t_k\}$) and the standard deviation of playing times ($\sigma\{t_k\}$) for all subjects are computed. We suggest that the $E\{t_k\}$ and $\sigma\{t_k\}$ values are appropriate measures to represent the level of challenge and the level of curiosity respectively [9] during game play. The former provides a notion for a goal whose attainment is uncertain (winning the game) — the lower the $E\{t_k\}$ value, the higher the goal uncertainty and furthermore the higher the challenge — and the latter effectively portrays a notion of unpredictability in the subsequent events of the game — the higher the $\sigma\{t_k\}$ value the more variable the game duration, so the higher the opponent unpredictability and therefore the higher the curiosity.

5.2 Tools

Two alternative neural network structures (a feedforward NN and a fuzzy-NN) for learning the relation between the challenge and curiosity factors and the entertainment value of a game have been used and are presented here. The assumption is that the entertainment value y of a given game is an unknown function of $E\{t_k\}$ and $\sigma\{t_k\}$, which the NN will learn. The subjects' expressed preferences constrain but do not specify the values of y for individual games. Since there is no *a priori* target y values for any given game, the output error function is not differentiable, and ANN training algorithms such as backpropagation are inapplicable. Learning is achieved through artificial evolution [34] and is described in Section 5.2.

Feedforward NN. A fully-connected multi-layered feedforward NN has been evolved [34] for the experiments presented here. The sigmoid function is employed at each neuron, the connection weights take values from -5 to 5 and both input values are normalized into $[0, 1]$ before they are entered into the feedforward NN. In an attempt to minimize the controller's size, it was determined that single hidden-layered NN architectures, containing 20 hidden neurons, are capable of successfully obtaining solutions of high fitness (network topology is not evolved, however).

Fuzzy-NN. A fuzzy [35] Sugeno-style [36] inference neural network is trained to develop fuzzy rules by evolving the memberships functions for both the input ($E\{t_k\}$, $\sigma\{t_k\}$) and the output variable y of the network as well as each fuzzy rule's weight. Each of the input and output values is presented by five fuzzy sets corresponding to very low, low, average, high and very high. The membership functions for the input values are triangular and their center α and width β are evolved whereas the output fuzzy sets use singleton membership functions [36] — only the center α of the spike membership function is evolved. The centroid technique is used as a defuzzification method.

Genetic Algorithm. A generational genetic algorithm (GA) [37] is implemented, which uses an "exogenous" evaluation function that promotes the minimization of the difference in matching the human judgement of entertainment. The feedforward NNs and fuzzy-NNs are themselves evolved. In the algorithm presented here, the evolving process is limited to the connection weights of the feedforward NN and the rule weights and membership function parameters of the fuzzy-NN.

The evolutionary procedure used can be described as follows. A population of N networks is initialized randomly. For feedforward NNs, initial real values that lie within $[-5, 5]$ for their connection weights are picked randomly from a uniform distribution, whereas for the fuzzy-NNs, initial rule weight values equal 0.5 and their membership function parameter values lie within $[0, 1]$ (uniformly distributed). Then, at each generation:

Step 1 Each member (neural network) of the population is evaluated with two pairs of ($E\{t_k\}$, $\sigma\{t_k\}$) values, one for A and one for B, and returns two output values, namely $y_{j,A}$ (interest value of the game set against opponent A) and $y_{j,B}$ (interest value of the game set against opponent B) for each pair j of sets played in the survey ($N_s = 30$). If the numerical relationship between $y_{j,A}$, $y_{j,B}$ matches the preference of subject j then we state that: 'the values agree with the subject' throughout this chapter (e.g. $y_{j,A} > y_{j,B}$ when subject j has a expressed a preference for A over B: $A \succ B$). In the opposite case, we state that: 'the values disagree with the subject.'

Step 2 Each member i of the population is evaluated via the fitness function f_i:

$$f_i = \sum_{j=1}^{N_s} \begin{cases} 1, & \text{if } y_{j,A}, y_{j,B} \text{ agree with subject } j; \\ \left(\frac{1-D(A,B)}{2}\right)^2, & \text{if } y_{j,A}, y_{j,B} \text{ disagree with subject } j. \end{cases} \quad (1.6)$$

where $D(A, B) = |y_{j,A} - y_{j,B}|$

Step 3 A fitness-proportional scheme is used as the selection method.

Step 4 Selected parents clone an equal number of offspring so that the total population reaches N members or reproduce offspring by crossover. The

Montana and Davis [38] crossover and the uniform crossover operator is applied for feedforward NNs and fuzzy-NNs respectively with a probability 0.4. Step 5 Gaussian mutation occurs in each gene (connection weight) of each offspring's genome with a small probability $p_m = 1/n$, where n is the number of genes.

The algorithm is terminated when either a good solution (i.e. $f_i > 29.0$) is found or a large number of generations g is completed ($g = 10000$).

5.3 Results

Results obtained from both feedforward NN and fuzzy-NN evolutionary approaches are presented in this section. In order to control for the non-deterministic effect of the GA initialization phase, each learning procedure (i.e. GA run) for each NN type is repeated ten times — we believe that this number is adequate to illustrate clearly the behavior of each mechanism — with different random initial conditions.

Evolved Feedforward NN. For space considerations, only the two fittest solutions achieved from the evolving feedforward NN approach are illustrated in Fig. 1. The qualitative features of the surfaces plotted in Fig. 1 appeared in all ten learning attempts. The most important conclusions derived from the feedforward NN mapping between $E\{t_k\}$, $\sigma\{t_k\}$ and entertainment are that:

- Entertainment has a low value when challenge is too high ($E\{t_k\} \approx 0$) and curiosity is low ($\sigma\{t_k\} \approx 0$).
- Even if curiosity is low, if challenge is at an appropriate level ($0.2 < E\{t_k\} < 0.6$), the game's entertainment value is high.
- If challenge is too low ($E\{t_k\} > 0.6$), the game's entertainment value appears to drop, independently of the level of curiosity.
- There is only a single data point present when $\sigma\{t_k\} > 0.8$ and the generalization of the evolved feedforward NNs within this space appears to be poor. Given that only one out of the 60 different game play data points falls in that region of the two-dimensional ($E\{t_k\}, \sigma\{t_k\}$) space, we can hypothesize that there is low probability for a game to generate curiosity values higher than 0.8. Thus, this region can be safely considered insignificant for these experiments. However, more samples taken from a larger game play survey would be required to effectively validate this hypothesis.

The fittest evolved feedforward NN is also tested against the custom-designed I metric for cross-validation purposes. The feedforward NN ranks the five different opponents previously mentioned in section 4.2 in the order $I_1 = I_2 < I_4 < I_3 < I_5$ (where I_i is the entertainment value the i opponent generates) which yields a correlation of 0.5432 (p-value = $3.89 \cdot 10^{-12}$) of agreement with human notion of entertainment expressed by the subject choices

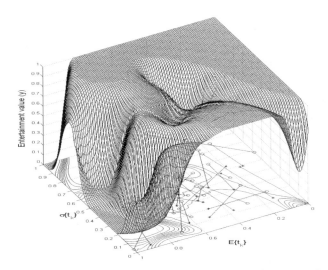

Fig. 1a. The fittest feedforward NN solution ($f = 29.95$)

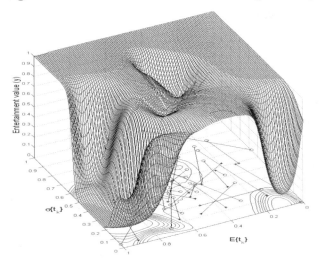

Fig. 1b. The second fittest feedforward NN solution ($f = 29.67$)

Fig. 1a–b. Circles ('o') and stars ('*') represent $E\{t_k\}$, $\sigma\{t_k\}$ values obtained by playing against opponents A and B respectively. Straight lines are used to connect the sets of games that humans played in pairs in the original experiment. Given this ranking of entertainment against these five opponents, the feedforward NN approach appears to model human entertainment better than the custom-designed interest metric proposed in [6, 22] and described above ($r = 0.4444$, p-value $= 1.17 \cdot 10^{-8}$).

The relationship between entertainment, challenge and curiosity expressed by the evolved feedforward NNs appears to follow the qualitative principles of Malone's work [9] and the human-verified interest metric developed in our previous work [6] for predator/prey games. According to these, a game should maintain an appropriate level of challenge and curiosity in order to be entertaining. In other words, too difficult and/or too easy and/or too unpredictable and/or too predictable opponents to play against make the game uninteresting.

5.4 Evolving Fuzzy-NN

The evolutionary procedure for the fuzzy-NN approach is also repeated ten times and only the fuzzy-NN that generates the highest fitness ($f = 29.81$) is presented here for reasons of space. Twenty five fuzzy rules are initially designed based the conclusions derived from the evolved feedforward NNs. The fittest fuzzy-NN generates 19 fuzzy rules in total — rules with weight values less than 0.1 are not considered significant and therefore are excluded from further consideration — which are presented here with their corresponding weight values w:

- Entertainment is *very low* if (a) challenge is *very high* and curiosity is *low* (Rule 1; $w_1 = 0.4440$) and (b) challenge is *low* and curiosity is *average* (Rule 2; $w_2 = 0.3617$).
- Entertainment is *low* if (a) challenge is *very low* and curiosity is *average* (Rule 3; $w_3 = 0.9897$) or *low* (Rule 4; $w_4 = 0.7068$); (b) challenge is *low* and curiosity is *high* (Rule 5; $w_5 = 0.7107$); (c) challenge is *high* and curiosity is *very low* (Rule 6; $w_6 = 0.5389$) and (d) challenge is *very high* and curiosity is *very low* (Rule 7; $w_7 = 0.9520$) or *high* (Rule 8; $w_8 = 0.9449$).
- Entertainment is *average* if challenge is *very low* and curiosity is *high* (Rule 9; $w_9 = 0.5818$).
- Entertainment is *high* if (a) challenge is *low* and curiosity is *very low* (Rule 10; $w_{10} = 0.8498$) or *very high* (Rule 11; $w_{11} = 0.2058$); (b) challenge is *average* and curiosity is *low* (Rule 12; $w_{12} = 0.5$); (c) challenge is *high* and curiosity is *low* (Rule 13; $w_{13} = 0.2824$) or *average* (Rule 14; $w_{14} = 0.25$) and (d) challenge is *very high* and curiosity is *average* (Rule 15; $w_15 = 0.2103$).
- Entertainment is *very high* if (a) challenge is *very low* and curiosity is *very high* (Rule 16; $w_{16} = 0.7386$); (b) challenge is *average* and curiosity is *very low* (Rule 17; $w_{17} = 0.5571$) or *very high* (Rule 18; $w_18 = 0.8364$) and (c) challenge is *high* and curiosity is *high* (Rule 19; $w_{19} = 0.2500$).

The quantitative measures of entertainment achieved through the neuro-fuzzy approach and the majority of the fuzzy rules generated appear consistent with Malone's principles of challenge and curiosity, the empirical contributions of the interest metric from the literature [22] and the fittest feedforward NN presented in section 5.3. However, the fittest fuzzy-NN (being less fit than the fittest feedforward NN) generates some few fuzzy rules that are not consistent with the aforementioned principles — e.g. Rule 10: entertainment is *high* if challenge is *low* and curiosity is *very low*. It is not clear whether the poorer performance is intrinsic to the method or a result of unlucky initialization; further tests are needed to distinguish these alternatives.

The fuzzy-NN is tested against the I metric of section 4 as in the evolved feedforward NN approach. The evolved fuzzy-NN ranks the five opponents in the order $I_2 < I_1 < I_3 < I_4 = I_5$. This ranking demonstrates a correlation of 0.3870 (p-value $= 1.74006 \cdot 10^{-6}$) of agreement with human notion of entertainment, which appears to be lower than the correlation achieved through the I value of section 4 ($r = 0.4444$, p-value $= 1.17 \cdot 10^{-8}$ [6]). However, as in the feedforward NN approach, the generalization of the evolved fuzzy-NNs appears to be poor when $\sigma\{t_k\} > 0.8$ due to the presence of a single data point within this region of the $(E\{t_k\}, \sigma\{t_k\})$ space. Even though we consider this non-frequent region as insignificant as far as this work is concerned, it may be sampled from a more extensive human game experiment in a future study.

5.5 Conclusions

This section introduced an alternative approach to constructing a quantitative metric of entertainment motivated by the qualitative principles of Malone's intrinsic factors for engaging game play [9]. More specifically, the quantitative impact of the factors of challenge and curiosity on human entertainment were investigated in the Pac-Man game.

The two neuro-evolution approaches for modeling entertainment examined demonstrate qualitative features that share principles with the interest metric (I value) presented in section 4. This (second) approach replaces the hand-crafted mathematical formulation of the interest metric with a more general process of machine learning applied to neural network models. Both obtained models manage to map successfully between the measures of entertainment factors such as challenge and curiosity and the notion of human game play satisfaction.

Validation results obtained show that the fittest feedforward NN gets closer — in the sense of statistical correlation — to the human notion of entertainment than both the I value [22] and the fittest fuzzy-NN. Therefore, it appears that the average and the standard deviation of a human's playing time over a number of games are in themselves adequate, and in fact more effective than the I value (as reported in [22]), for capturing player entertainment in real-time in predator/prey games.

The reported work on this approach is most significantly limited by the number of participants in the game survey we devised. Therefore, not all regions of the challenge-curiosity search space were sampled by human play which therefore yielded poor NN generalization for these regions. Limited data also restricted the sensible number of inputs to the learning system.

Malone's entertainment factor of fantasy is omitted here since the focus is on the contribution of the opponent behaviors to the generation of entertainment; however, experiments on interactive physical predator/prey games with children have shown that entertainment increases monotonically with respect to the fantasy factor [39].

This second entertainment modeling approach presented here demonstrates generality over the majority of computer game genres since the quantitative means of challenge and curiosity are estimated through a generic feature of game play which is the playing time of humans over a number of games. Thus, these or similar measures could be used to measure player satisfaction in any genre of game. However, each game possesses additional idiosyncratic entertainment features that might need to be extracted and added to the proposed generic measures used as input to the machine learning tools — therefore, more games of the same or other genres need to be tested to evaluate the true generality of this approach.

6 Discussion

The foregoing has proposed and demonstrated a pair of methods for deriving a quantitative estimate of the level of entertainment experienced by a player of a computer game, using data that can be derived from or during game play. In this section, we discuss some of the questions raised by the approach and the assumptions on which it is based.

An immediate and natural question is whether the techniques described really capture "entertainment" which, after all, is a complex mental phenomenon depending on the player, the game and (probably) a number of external factors in rather involved ways. We acknowledge that the definition of a quantitative metric for entertainment in this sense is almost certainly infeasible. However, we take a practical approach here: it is sufficient for our purposes if a quantity exists that can be computed from observable data from the player-game interaction and that correlates well with players' expressed preferences. In other words, a numerical value which orders games in the same way as players' judgement of entertainment is sufficient for our purposes.

The foregoing material illustrates two ways to construct such a metric: by design, using the insights of a skilled game player; and by using machine learning to explore the space of possible evaluation functions whose values are consistent with human judgement of entertainment value. The resulting metric is specific to the particular game under consideration, but the general method of construction is applicable to a wide variety of game instances and genres.

To summarize, therefore, the proposed approach does not capture details of the complex mental states associated with enjoyment of computer games, but it does provide a way to evaluate different instances of game play in terms of how entertaining they are for a player. Such knowledge can be used, for example, for tuning the game to suit the player (see below).

In addition to this general question concerning the approach, there are a number of assumptions (and hence limitations) associated with the methods presented; these are discussed below.

6.1 Assumptions and Limitations of the I Value

The interest metric described in section 4 is based on specific assumptions about the features of a game that generate enjoyment for the player.

- The approach assumes that interaction with opponent behavior is the primary source of variability in the entertainment value of a given game. That is, the enjoyment generated by the graphical, multimedia and storyline components of the game design is disregarded. This is a reasonable assumption for comparisons between instances of play of a single given game, but means that the interest metric is specific to a certain game and cannot be used for meaningful comparison between different games (e.g. to answer "Is Space Invaders more entertaining than Quake?").
- The interest metric is calculated using data obtained from a sample of N games. This is consistent with human cognition since it appears that human players require a significant number of games (or else playing time) to classify a specific computer game according to their perceived satisfaction. However, this assumption constitutes a limitation of the method. A further investigation of the relationship between the I value and the N played games might reveal that fewer games are needed for an estimate that is still consistent with human notion of perceived entertainment.
- The interest value definition assumes that players of the game have average-level playing skills. By 'average-level' we only exclude the two following extreme player types: (1) those who have never played the specific game before and furthermore do not know the rules and how to play it — these players perform poorly against almost any type of opponent; (2) those who have an excellent knowledge of the specific game, can easily predict the opponent behavior and have mastered the game controls. These players can usually beat even the most effective opponents designed for the game. In each case, the interest value might not be very well estimated since the challenge criterion T approximates a zero value regardless of the opponent, in the first case because the game is much too hard and in the second because it is too easy (recall that T is designed to

be maximum for a 'reasonably difficult' game). This appears to be consistent with human notion of interestingness since we believe that neither extreme player type will find the game interesting.
- The interest value depends primarily on the opponents' behavior. Implicitly, through that, it depends on the player's behavior since the opponent behavior is elicited in interaction with the player. (The previous point concerning playing skills is a specific instance of this more general observation). If the players of a game can be divided into classes with quite different playing styles, for example "aggressive" vs. "defensive", then it may be necessary to design a separate I metric formula for each player type, because of the difference in opponent behavior elicited by their differing styles of play. For a comprehensive discussion of this assumption in [32] where the interest value dependence on the player is investigated through experiments in the Pac-Man test-bed game.
- A factor that may contribute to enjoyable games is the match between the real-time speed of the game and the reaction time of the player. Gradually decreasing the required player reaction time is a standard and inexpensive technique used by a set of games (i.e Pac-Man) to achieve increasing challenge for the player as the game proceeds. This is not considered in the work in this chapter since changing the demanded reaction time does not alter the qualitative properties of the opponent behavior (except through the implicit dependence on the player's style of play). Note that in the extreme case, an unintelligent opponent may generate an exciting game just because of the unrealistically fast reaction time required of the player.

6.2 Assumptions and Limitations of the Machine Learning Approach

The second approach to constructing a metric uses machine learning rather than designer insight to build a quantitative evaluation of a game. This method is based on the same fundamental assumptions as the I value approach: that the opponents' behavior is the primary determinant of the entertainment value of a given instance of game play. Many of the comments of the previous section also apply to this approach. However, there are a few observations specific to this methodology.

- The issue of playing style is arguably less relevant here than in the designer-insight method. If it is necessary to determine player type to be able to evaluate games, then the requirement of a consistent metric will in principle force the machine learning technique to perform an implicit player-type classification, assuming that the function representation technology (here, a neural network) is powerful enough to do so. In other words, because this approach can construct much more complex (and therefore less scrutable) mappings from raw measurements to entertainment metric value, it can cope with more complex relationships between the data and the metric than a designer is likely to be able to manage.

- However, the effectiveness of a machine learning method depends strongly on the quantity and quality of data available. For the case considered here, this data comprises two kinds: raw measurements derived from game play, that represent aspects of features such as challenge, curiosity, fantasy, etc.; and expressed preferences or rankings between instances of game play. The former provide the observables on which the metric is built, the latter determine the degree of consistency with human judgement of any given proposal for the metric function.
  Obtaining the latter kind of data involves experimentation in which players are asked to say which of several (here two) instances of game play they prefer; collecting such data is time- and player-consuming. This limits the complexity of metric that can be considered, since it limits the feedback available to the machine learning process during the exploration of the space of metrics.
- The former kind of data also presents certain problems: specifically, how do we know what measurements to include as a basis for the metric? The work presented here uses one designer-chosen measurement for each relevant feature — challenge and curiosity — but one could in principle devise many measurements correlated with either feature, and allow the machine learning system to use all of them or to make a selection of the best measurements for the purpose. This approach removes a certain designer bias in the method at the expense of complicating the machine learning task and approaching earlier the limits imposed by the difficulty of collecting preference data from players.
- The issue of what value of N to choose can be finessed in this second approach, as can the question of game speed and demanded player reaction time, by appropriate choice of measurements from which to build the evaluation metric. For instance, game speed can be included directly as a measurement and measurements requiring different numbers of games to compute can be included in the process.

6.3 Making Use of Entertainment Metrics

Given a successful metric of entertainment for a given game, designed and evaluated using one of the methods proposed above, the final question we consider here is how such knowledge might be used. As previously noted, opponents which can learn and adapt to new playing strategies offer a richer interaction to entertain the player. An obvious use of the entertainment metric is therefore to adapt the game so that the metric value increases.

Two possible strategies for this might be:

- to use a machine learning mechanism for the game studied which allows opponents to learn while playing against the player (i.e. on-line). The entertainment metric can be used to guide the learning process.

Such an on-line learning mechanism is comprehensively described in [6, 33, 22]. Its basic steps are presented briefly here as follows. At each generation of the algorithm:

Step 1: Each opponent is evaluated every e_p simulation steps via an individual reward function that provides an estimate of the I value of the game, while the game is played.

Step 2: A pure elitism selection method is used where only a small percentage of the fittest opponents is able to breed. The fittest parents clone offspring.

Step 3: Mutation occurs in each opponent (varying neural network controller connection weights) with a suitable small probability.

Step 4: Each mutated offspring is evaluated briefly in off-line mode, that is, by replacing the least-fit member of the population and playing a short off-line game of e_p simulation steps against a selected computer-programmed player. The fitness values of the mutated offspring and the least-fit member are compared and the better one is deployed in the game.

The algorithm is terminated when a predetermined number of games has been played or a game of high interest (e.g. $I \geq 0.7$) is found.

Results reported in [21, 22] demonstrate that 50 on-line learning games are not enough (in Pac-Man) for the on-line learning mechanism to cause a difference in the I value which is noticeable by human players. The on-line learning period of 50 games is an empirical choice to balance efficiency and experimental time. The duration of the on-line learning procedure in this experiment lasted 20 minutes on average while the whole human survey experiment presented in section 4.2 and section 5.1 exceeded 65 minutes in many cases, which is a great amount of time for a human to concentrate.

– to use a metric evaluation function constructed using the machine learning technique directly to enhance the entertainment provided by the game. The key to this is the observation that the models (feedforward NN or fuzzy-NN) relate game features to entertainment value. It is therefore possible in principle to infer what changes to game features will cause an increase in the interestingness of the game, and to adjust game parameters to make those changes. For the feedforward NN, the partial derivatives of $\vartheta y/\vartheta E\{t_k\}$ and $\vartheta y/\vartheta \sigma\{t_k\}$ indicate the change in entertainment for a small change in an individual game feature. One could use gradient ascent to attempt to improve entertainment with such a model. The fuzzy-NN approach provides qualitative rules relating game features to entertainment, rather than a quantitative function, but an analogous process could be applied to augment game entertainment.

Such a direction constitutes an example of future work within computer, physical and educational games. The level of engagement or motivation of the user/player/gamer of such interactive environments can be increased by the use of the presented approaches providing systems of richer interaction and qualitative entertainment [5],

Acknowledgements

This work was supported in part by the Danish Research Agency, Ministry of Science, Technology and Innovation (project no: 274-05-0511).

Resources

Key Books

- Koster, R., A Theory of Fun for Game Design, Paraglygh Press, 2005.

Key Papers

- Andrade, G., Ramalho, G., Santana, H., Corruble, V. "Challenge-Sensitive Action Selection: an Application to Game Balancing," in *Proceedings of the IEEE/WIC/ACM International Conference on Intelligent Agent Technology (IAT-05)*, pp. 194–200, Compiegne, France. IEEE Computer Society, 2005.
- Iida, H., Takeshita, N., Yoshimura, J., "A metric for entertainment of boardgames: its implication for evolution of chess variants," in *Nakatsu, R., Hoshino, J., eds.: IWEC2002 Proceedings*, pp. 65–72, Kluwer, 2003.
- Malone, T. W., "What makes computer games fun?," *Byte*, vol. 6, pp. 258–277, 1981.
- Lazzaro, N., "Why we play games: Four keys to more emotion without story," Technical report, XEO Design Inc. 2004.
- Yannakakis, G. N., Hallam, J., "Evolving Opponents for Interesting Interactive Computer Games," in *Proceedings of the 8th International Conference on the Simulation of Adaptive Behavior (SAB'04); From Animals to Animats 8*, pp. 499–508, Los Angeles, CA, USA, July 13–17, 2004. The MIT Press.
- Yannakakis, G. N., Hallam, J., "Towards Optimizing Entertainment in Computer Games," *Applied Artificial Intelligence*, 2007 (to appear).

Discussion Groups, Forums

- Optimizing player satisfaction in games discussion group:
  http://groups.google.com/group/optimizing-player-satisfaction-in-games
- Player experience special interest group of DIGRA; list url:
  http://mail.digra.org/mailman/listinfo/player-experience

Key International Conferences/Workshops

Workshop series on Optimizing Player Satisfaction. In conjunction with
- Simulation of Adaptive Behaviour (SAB) Conference, Rome, Italy, in 2006 and
- Artificial Intelligence and Interactive Digital Entertainment (AIIDE) Conference, Stanford, US, in 2007.

References

1. Champandard, A.J.: AI Game Development. New Riders Publishing (2004)
2. Funge, J.D.: Artificial Intelligence for Computer Games. A. K. Peters Ltd, (Wellesley, Massachusetts, USA)
3. Laird, J.E., van Lent, M.: Human-level AI's killer application: Interactive computer games. In: Proceedings of the Seventh National Conference on Artificial Intelligence (AAAI). (2000) 1171–1178
4. Vorderer, P., Hartmann, T., Klimmt, C.: Explaining the enjoyment of playing video games: the role of competition. In Marinelli, D., ed.: ICEC conference proceedings 2003: Essays on the future of interactive entertainment, Pittsburgh, (Carnegie Mellon University Press) 107–120
5. Yannakakis, G.N., Hallam, J.: A scheme for creating digital entertainment with substance. In: Proceedings of the Workshop on Reasoning, Representation, and Learning in Computer Games, 19th International Joint Conference on Artificial Intelligence (IJCAI). (2005) 119–124
6. Yannakakis, G.N., Hallam, J.: Evolving Opponents for Interesting Interactive Computer Games. In Schaal, S., Ijspeert, A., Billard, A., Vijayakumar, S., Hallam, J., Meyer, J.A., eds.: From Animals to Animats 8: Proceedings of the 8th International Conference on Simulation of Adaptive Behavior (SAB-04), Santa Monica, LA, CA, The MIT Press (2004) 499–508
7. Koster, R.: A Theory of Fun for Game Design. Paraglyph Press (2005)
8. Kapoor, A., Mota, S., Picard, R.: Towards a Learning Companion that Recognizes Affect. In: Proceedings of Emotional and Intelligent II: The Tangled Knot of Social Cognition, AAAI Fall Symposium (2001)
9. Malone, T.W.: What makes computer games fun? Byte **6** (1981) 258–277
10. Csikszentmihalyi, M.: Flow: The Psychology of Optimal Experience. New York: Harper & Row (1990)
11. Sweetser, P., Wyeth, P.: GameFlow: A Model for Evaluating Player Enjoyment in Games. ACM Computers in Entertainment **3** (2005)
12. Lazzaro, N.: Why we play games: Four keys to more emotion without story. Technical report, XEO Design Inc. (2004)
13. Read, J., MacFarlane, S., Cassey, C.: Endurability, engagement and expectations. In: Proceedings of International Conference for Interaction Design and Children. (2002)
14. Kline, S., Arlidge, A.: Online Gaming as Emergent Social Media: A Survey. Technical report, Media Analysis Laboratory, Simon Fraser University (2003)
15. Crispini, N.: Considering the growing popularity of online games: What contibutes to making an online game attractive, addictive and compeling. Dissertation, SAE Institute, London (2003)

16. Choi, D., Kim, H., Kim, J.: Toward the construction of fun computer games: Differences in the views of developers and players. Personal Technologies **3** (1999) 92–104
17. Yannakakis, G.N., Lund, H.H., Hallam, J.: Modeling Children's Entertainment in the Playware Playground. In: Proceedings of the IEEE Symposium on Computational Intelligence and Games, Reno, USA, IEEE (2006) 134–141
18. Nareyek, A.: Intelligent agents for computer games. In Marsland, T., Frank, I., eds.: Computers and Games, Second International Conference, CG 2002. (2002) 414–422
19. Taatgen, N.A., van Oploo, M., Braaksma, J., Niemantsverdriet, J.: How to construct a believable opponent using cognitive modeling in the game of set. In: Proceedings of the fifth international conference on cognitive modeling. (2003) 201–206
20. Iida, H., Takeshita, N., Yoshimura, J.: A metric for entertainment of boardgames: its implication for evolution of chess variants. In Nakatsu, R., Hoshino, J., eds.: IWEC2002 Proceedings, Kluwer (2003) 65–72
21. Yannakakis, G.N.: AI in Computer Games: Generating Interesting Interactive Opponents by the use of Evolutionary Computation. Ph.d. thesis, University of Edinburgh (2005)
22. Yannakakis, G.N., Hallam, J.: Towards Optimizing Entertainment in Computer Games. Applied Artificial Intelligence (2007) to appear.
23. Yannakakis, G.N., Hallam, J.: Towards Capturing and Enhancing Entertainment in Computer Games. In: Proceedings of the 4th Hellenic Conference on Artificial Intelligence, Lecture Notes in Artificial Intelligence. Volume 3955., Heraklion, Greece, Springer-Verlag (2006) 432–442
24. Andrade, G., Ramalho, G., Santana, H., Corruble, V.: Extending reinforcement learning to provide dynamic game balancing. In: Proceedings of the Workshop on Reasoning, Representation, and Learning in Computer Games, 19th International Joint Conference on Artificial Intelligence (IJCAI). (2005) 7–12
25. Verma, M.A., McOwan, P.W.: An adaptive methodology for synthesising Mobile Phone Games using Genetic Algorithms. In: Congress on Evolutionary Computation (CEC-05), Edinburgh, UK (2005) 528–535
26. Hunicke, R., Chapman, V.: AI for Dynamic Difficulty Adjustment in Games. In: Proceedings of the Challenges in Game AI Workshop, 19th Nineteenth National Conference on Artificial Intelligence (AAAI'04). (2004)
27. Spronck, P., Sprinkhuizen-Kuyper, I., Postma, E.: Difficulty Scaling of Game AI. In: Proceedings of the 5th International Conference on Intelligent Games and Simulation (GAME-ON 2004). (2004) 33–37
28. Yannakakis, G.N., Hallam, J., Lund, H.H.: Capturing Entertainment through Heart-rate Dynamics in the Playware Playground. In: Proceedings of the 5th International Conference on Entertainment Computing, Lecture Notes in Computer Science. Volume 4161., Cambridge, UK, Springer-Verlag (2006) 314–317
29. Rani, P., Sarkar, N., Liu, C.: Maintaining optimal challenge in computer games through real-time physiological feedback. In: Proceedings of the 11th International Conference on Human Computer Interaction. (2005)
30. McQuiggan, S., Lee, S., Lester, J.: Predicting User Physiological Response for Interactive Environments: An Inductive Approach. In: Proceedings of the 2nd Artificial Intelligence for Interactive Digital Entertainment Conference. (2006) 60–65

31. Yannakakis, G.N., Hallam, J.: A generic approach for generating interesting interactive pac-man opponents. In Kendall, G., Lucas, S., eds.: Proceedings of the IEEE Symposium on Computational Intelligence and Games, Essex University, Colchester, UK (2005) 94–101
32. Yannakakis, G.N., Maragoudakis, M.: Player modeling impact on player's entertainment in computer games. In: Proceedings of the 10th International Conference on User Modeling; Lecture Notes in Computer Science. Volume 3538., Edinburgh, Springer-Verlag (2005) 74–78
33. Yannakakis, G.N., Hallam, J.: A Generic Approach for Obtaining Higher Entertainment in Predator/Prey Computer Games. Journal of Game Development **1** (2005) 23–50
34. Yao, X.: Evolving artificial neural networks. In: Proceedings of the IEEE. Volume 87. (1999) 1423–1447
35. Zadeh, L.: Fuzzy sets. Information and Control **8** (1965) 338–353
36. Sugeno, M.: Industrial Applications of Fuzzy Control. North-Holland (1985)
37. Holland, J.H.: Adaptation in Natural and Artificial Systems. University of Michigan Press, Ann Arbor, MI (1975)
38. Montana, D.J., Davis, L.D.: Training feedforward neural networks using genetic algorithms. In: Proceedings of the Eleventh International Joint Conference on Artificial Intelligence (IJCAI-89), San Mateo, CA, Morgan Kauffman (1989) 762–767
39. Yannakakis, G.N., Hallam, J., Lund, H.H.: Comparative Fun Analysis in the Innovative Playware Game Platform. In: Proceedings of the 1st World Conference for Fun 'n Games. (2006) 64–70